Advanced materials in automotive engineering

Related titles:

Diesel engine system design
(ISBN 978-1-84569-715-0)
Diesel engine system design links everything diesel engineers need to know about engine performance and system design in order for them to master all the essential topics quickly and to solve practical design problems. Based on the author's unique experience in the field, it enables engineers to come up with an appropriate specification at an early stage in the product development cycle.

Tailor welded blanks for advanced manufacturing
(ISBN 978-1-84569-704-4)
Tailor welded blanks are sheets made from different strengths and thicknesses of steel pre-welded together being pressed and shaped into the final component. They produce high-quality components with the right grade and thickness of steel where they are most needed, providing significant savings in weight and processing costs in such industries as automotive engineering. Part I reviews processing issues in product design, production methods, weld integrity and deformation. Part II discusses applications in areas such as automotive and aerospace engineering.

Handbook of metal injection molding
(ISBN 978-0-85709-066-9)
Metal injection molding (MIM) is an important technology for the manufacture of small and intricate components with a high level of precision. MIM components are used in sectors such as automotive and biomedical engineering as well as microelectronics. This book is an authoritative guide to the technology and its applications. The book reviews key processing technologies, quality issues and MIM processing of a range of metals.

Details of these and other Woodhead Publishing materials books can be obtained by:

- visiting our web site at www.woodheadpublishing.com
- contacting Customer Services (e-mail: sales@woodheadpublishing.com; fax: +44 (0) 1223 832819; tel.: +44 (0) 1223 499140 ext. 130; address: Woodhead Publishing Limited, 80 High Street, Sawston, Cambridge CB22 3HJ, UK)
- contacting our US office (e-mail: usmarketing@woodheadpublishing.com; tel.: (215) 928 9112; address: Woodhead Publishing, 1518 Walnut Street, Suite 1100, Philadelphia, PA 19102-3406, USA)

If you would like e-versions of our content, please visit our online platform: www.woodheadpublishingonline.com. Please recommend it to your librarian so that everyone in your institution can benefit from the wealth of content on the site.

Advanced materials in automotive engineering

Edited by

Jason Rowe

WOODHEAD
PUBLISHING

Oxford   Cambridge   Philadelphia   New Delhi

Published by Woodhead Publishing Limited,
80 High Street, Sawston, Cambridge CB22 3HJ, UK
www.woodheadpublishing.com
www.woodheadpublishingonline.com

Woodhead Publishing, 1518 Walnut Street, Suite 1100, Philadelphia,
PA 19102-3406, USA

Woodhead Publishing India Private Limited, G-2, Vardaan House,
7/28 Ansari Road, Daryaganj, New Delhi – 110002, India
www.woodheadpublishingindia.com

First published 2012, Woodhead Publishing Limited
© Woodhead Publishing Limited, 2012
The authors have asserted their moral rights.

British Library Cataloguing in Publication Data
A catalogue record for this book is available from the British Library.

Library of Congress Control Number: 2012931665

ISBN 978-1-84569-561-3 (print)
ISBN 978-0-85709-546-6 (online)

The publisher's policy is to use permanent paper from mills that operate a sustainable forestry policy, and which has been manufactured from pulp which is processed using acid-free and elemental chlorine-free practices. Furthermore, the publisher ensures that the text paper and cover board used have met acceptable environmental accreditation standards.

Typeset by Replika Press Pvt Ltd, India
Printed by Lightning Source

Contents

Contributor contact details

(* = main contact)

Editor and Chapter 1

J. Rowe

E-mail: rowejmc@gmail.com

Chapter 2

P. K. Mallick
Department of Mechanical
    Engineering
University of Michigan – Dearborn
4901 Evergreen Road
Dearborn, MI 48128
USA

E-mail: pkm@umich.edu

Chapter 3

Nick den Uijl* and Louisa Carless
Tata Steel RD&T
P.O. Box 10.000
1970 CA Ijmuiden
The Netherlands

E-mail: nick.den-uijl@tatasteel.com;
    louisa.carless@tatasteel.com

Chapter 4

Y. Okitsu*
Honda R&D Co. Ltd
4930 Shimotakanezawa
Haga-machi, Haga-gun
Tochigi 321-3393
Japan

E-mail: yoshitaka_okitsu@n.t.rd.honda.
    co.jp

N. Tsuji
Department of Materials Science
    and Engineering
Graduate School of Engineering
Kyoto University
Yoshida-Honmachi, Sakyo-ku
Kyoto 606-8501
Japan

E-mail: nobuhiro.tsuji@ky5.ecs.kyoto-u.
    ac.jp

Chapter 5

M. Bloeck
Novelis Switzerland SA
Research and Development Centre
    Sierre
CH – 3960 Sierre
Switzerland

E-mail: margarete.bloeck@novelis.com

Chapter 6

F. Casarotto*, A. J. Franke and R.
    Franke
Rheinfelden Alloys GmbH & Co.
    KG
Friedrichstrasse 80
79618 Rheinfelden
Germany

E-mail: fcasarotto@rheinfelden-alloys.
    eu; franke@alurheinfelden.com;
    rfranke@rheinfelden-alloys.eu

Chapter 7

B. R. Powell*
Materials Battery Group
Electrochemical Energy Research
    Lab
Mail Code 480-102-000
General Motors Global Research
    and Development Center
30500 Mound Road
Warren, MI 48090-9055
USA

E-mail: bob.r.powell@gm.com

A.A. Luo
Light Metals for Powertrain and
    Structural Subsystems Group
    Chemical Sciences and Materials
    Systems Lab
Mail Code 480-106-212
General Motors Global Research
    and Development Center
30500 Mound Road
Warren, MI 48090-9055
USA

E-mail: alan.luo@gm.com

P. E. Krajewski
Front and Rear Closures Group
Mail Code 480-210-2Y9
General Motors Global Vehicle
    Engineering
30001 Van Dyke Road
Warren, MI 48090-9020
USA

E-mail: paul.e.krajewski@gm.com

Chapter 8

P. Mitschang* and K. Hildebrandt
Department of Manufacturing
    Science
Institut für Verbundwerkstoffe
    GmbH
Erwin-Schrödinger-Strasse, Geb. 58
67663 Kaiserslautern
Germany

E-mail: peter.mitschang@ivw.uni-kl.de

Chapter 9

P. Urban* and R. Wohlecker
Forschungsgesellschaft
    Kraftfahrwesen mbH Aachen
Steinbachstrasse 7
52074 Aachen
Germany

E-mail: urban@fka.de; wohlecker@
    fka.de

Chapter 10

T. Bein*, J. Bös, D. Mayer and T.
    Melz
Fraunhofer Institute for Structural
    Durability and System
    Reliability LBF
Bartningstrasse 47
64289 Darmstadt
Germany

E-mail: thilo.bein@lbf.fraunhofer.de

Chapter 11

K. Kirwan* and B. M. Wood
WMG
International Manufacturing Centre
University of Warwick
Coventry CV4 7AL
UK

E-mail: Kerry.Kirwan@warwick.ac.uk;
    b.m.wood@warwick.ac.uk

Chapter 12

J. A. Poulis* and F.M. De Wit
Delft University of Technology
Building 62
Kluyverweg 1
2629HS Delft
The Netherlands

E-mail: J.A.Poulis@tudelft.nl;
    F.M.deWit@tudelft.nl

Introduction: advanced materials and vehicle lightweighting

J. ROWE, Automotive Consultant Engineer, UK

The UK automotive industry is a large and critical sector within the UK economy. It accounts for 820,000 jobs, exports finished goods worth £8.9bn annually and adds value of £10 billion to the UK economy each year [1]. However, the UK automotive industry is currently facing great challenges as road transport released 132 million tonnes CO_2 in 2008, accounting for 19% of the total UK annual CO_2 emission. Furthermore, its global competitiveness is threatened by the emerging new economic powers, such as China and India. In addition, the UK government is committed to reduce CO_2 significantly by 2050 and the EU requires 95% recovery and reuse of ELVs (end of life vehicles) by 2015. A solution to these challenges comes from the development and manufacture of LCVs (low carbon vehicles), and this is clearly presented in the vision of the UK automotive industry set by the NAIGT [1].

Vehicle lightweighting is an effective approach to improve fuel economy and reduce CO_2 emissions. CO_2 emission per km driven is linearly related to vehicle curb weight [2]. Studies have shown that every 10% reduction in vehicle weight can result in 3.5% improvement in fuel efficiency (on the New European Drive Cycle (NEDC)) [3]. In terms of greenhouse effect, this means that every 100kg weight reduction results in CO_2 reduction of about $3.5gCO_2$/km driven for the entire vehicle life [3]. In addition to such primary benefits, vehicle lightweighting reduces the power required for acceleration and braking, which provides the opportunity to employ smaller engines, and smaller transmissions and braking systems. These savings have been termed secondary weight reduction in the literature and would allow a CO_2 reduction of up to $8.5gCO_2$/km [3]. Furthermore, if appropriate technologies are used, vehicle weight reduction can be achieved independent of size, functionality and class of vehicle. It is important to point out that similar benefits of mass reduction can be demonstrated for hybrid vehicles (HVs) and electric vehicles (EVs).

Approaches to vehicle mass reduction include deployment of advanced materials and mass-optimised vehicle design. One of the major systems of the vehicle is the body (body-in-white, or BIW) that represents about one-quarter of the overall vehicle mass and is the core structure and frame of the vehicle. The body is so fundamental to the vehicle that sometimes it is the only portion of the vehicle that is researched, designed and analysed in mass reduction technology studies [2]. Over many years there has been a fundamental material shift from wood, cast iron and steel to high strength steel (HSS), advanced high strength steel (AHSS), aluminium, magnesium and polymer matrix composites (PMCs). Between 1995 and 2007, the use of aluminium increased by 23%, PMCs by 25% and magnesium by 127% [2]. Further vehicle mass reduction can be achieved by mass-optimised design technology. Mass-optimisation from a whole vehicle perspective opens up the possibility for much larger vehicle mass reduction. For example, secondary mass reduction is possible since reducing the mass of one vehicle part can lead to further reductions elsewhere due to reduced requirements of the powertrain, suspension and body structure to support and propel the various systems. New and more holistic approaches that include integrated vehicle system design, secondary mass effects, multi-materials concepts and new manufacturing processes are expected to contribute to vehicle mass optimisation for much greater potential mass reduction [4]. As reviewed by Lutsey [2], there have been 26 major R&D programmes worldwide on vehicle mass reduction. Compared to a steel structure, the HSS intensive body structure by the Auto Steel Partnership achieved 20–30% mass reduction [5], the Al intensive body structures of the Jaguar XJ, Audi A8 and A2 achieved 30–40% mass reduction (e.g. [6]) and a multi-material body structure featuring more Al (37%), Mg (30%) and PMCs (21%) by the Lotus High Development Programme achieved 42% mass reduction [4]. It is clear that although a single material approach can achieve substantial mass reduction the greatest potential comes from an integrated multi-material approach that exploits the mass and functional properties of Al, Mg, PMCs and AHSS. Despite the greater use of the higher cost advanced materials, mass-optimised vehicle designs could have a minimal or moderate cost impact on new vehicles [2] if a holistical whole vehicle design approach is used. For instance, the Lotus High Development Programme demonstrated a 30% whole vehicle mass reduction could be achieved with only a 5% increase in cost, whilst the VW-led Super Light Car achieved a 35% body mass reduction for a cost of less than €8 for every kilogram of mass reduction. The combination of a multi-material concept and a mass-optimised whole vehicle design approach can achieve significant mass reduction with a minimal or moderate cost impact on vehicle structure and it is most likely that the future materials for LCVs are an optimised combination of Al, Mg, PMCs and AHSS.

Closed-loop recycling of advanced automotive materials, however, has been missing from nearly all the LCV programmes worldwide, which have concentrated on the reduction of CO_2 emission during the use phase of vehicles produced from primary advanced materials. The production energy of all primary automotive materials is always much greater than that of their secondary (recycled) counterparts [7]. For instance, production of 1kg primary Al from the primary route costs 45kWh electricity and releases 12kg

CO_2, whilst 1 kg recycled Al only costs only 5% of that energy and 5% CO_2 emission [8]. Detailed life cycle analysis (LCA) has shown that a primary Al intensive car can only achieve energy saving after more than 20,000 km driven compared with its steel counterpart, while a secondary Al intensive car will save energy from the very beginning of vehicle life [9]. If all the automotive materials can be effectively recycled in a closed-loop through advanced materials development and novel manufacturing technologies, the energy savings and cost reduction for the vehicle structure will be considerably more significant.

The vision of automotive manufacturers is that future LCVs are achieved by a combination of multi-material concepts with mass-optimised design approaches through the deployment of advanced low carbon input materials, efficient low carbon manufacturing processes and closed-loop recycling of ELVs. Advanced materials will include Al, Mg and PMCs, which are all supplied from a recycled source. A holistic and systematic mass-optimised design approach will be used throughout the vehicle (including chassis, trim, etc.) not only for mass reduction and optimised performance during vehicle life but also for facilitating reuse, remanufacture and closed-loop recycling at the end of vehicle life. Novel manufacturing processes will be used to reduce materials waste and energy consumption, shorten manufacturing steps and facilitate parts integration and ELV recycling. Fully closed-loop ELV recycling will be facilitated by new materials development, novel design approaches, advanced manufacturing processes and efficient disassembly technologies, all of which will be effectively guided by a full life cycle analysis.

The themes described above have been taken from the TARF-LCV 2011 (Towards Affordable, Closed-Loop Recyclable Future Low Carbon Vehicles Structures) programme submission (reproduced with the kind permission of Professor Zhongyun Fan, Chair of Metallurgy at Brunel University), and are developed within the following chapters of this book using contributions from leading experts from both academia and industry.

1.1   References

[1] NAIGT: *An Independent Report on the Future of the Automotive Industry in the UK*, 2009.
[2] N. Lutsey: UCD-ITS-RR-10-10, University of California, Davis, May 2010.
[3] M. Goede: SLC Project December 2003 and May 2008, VW.
[4] Lotus Engineering Inc: An Assessment of Mass Reduction Opportunities for a 2017-2020 Model Year Vehicle Programme, March 2010.
[5] Auto Steel Partnership (ASP): Future Generation of Passenger Compartment, December 2007.
[6] S. Birch: Jaguar Remakes XJ, http://www.sae.org/mags/sve/7547, March 2010.

[7] S. K. Das: in *Materials, Design and Manufacturing for Lightweight Vehicles*, CRC Press, New York, 2010, pp. 309–331.

[8] J. A. S. Green: Aluminium Recycling and Processing, ASM International, Materials Park, Ohio, USA, 2007, p. 67.

[9] T. Inaba: in *Automotive Engineering – Lightweight, Functional and Novel Materials*, Taylor and Francis, 2008, pp. 19–27.

2
Advanced materials for automotive applications: an overview

P. K. MALLICK, University of Michigan – Dearborn, USA

Abstract: With increasing demand on fuel economy improvement and emission control, there is great deal of interest in using advanced materials to produce lightweight vehicles. The advanced materials include advanced high strength steels, non-ferrous alloys, such as aluminum, magnesium and titanium alloys, and a variety of composites, including carbon fiber composites, metal matrix composites and nanocomposites. This chapter provides an overview of these materials and their current applications and potential applications in future automobiles.

Key words: advanced materials, advanced steels, aluminum alloys, magnesium alloys, titanium alloys, stainless steels, cast iron, composites, glazing materials.

2.1   Introduction

Vehicle weight reduction through material substitution is one of the key elements in the overall strategy for fuel economy improvement and emission control. While the principal material in current vehicles is plain carbon steels, there is now a great deal of interest in replacing them with advanced high strength steels, light non-ferrous alloys, such as aluminum, magnesium and titanium alloys, and a variety of composites, including carbon fiber composites, metal matrix composites and nanocomposites. This chapter is a broad overview of these materials and their applications in the automobiles.

2.1.1   Materials scenario

Plain carbon steel and cast iron were the workhorse materials in the automotive industry prior to 1970s. As shown in Table 2.1, even today steel is used in much larger quantities than any other material; however, high strength steels and advanced high strength steels, on account of their significantly higher strength, are now replacing plain carbon steels in several body structure and chassis applications. As a result, the amount of high strength and advanced high strength steels has increased in recent years, while the amount of plain carbon steels has decreased (Table 2.2). There is also an increasing use of aluminum alloys and polymer matrix composites. For example, the use of aluminum alloys in North American automobiles has increased from 2% of

5

Table 2.1 Material distribution in typical automobiles

Material	Percentage of vehicle weight	Major areas of application
Steel	55	Body structure, body panels, engine and transmission components, suspension components, driveline components
Cast iron	9	Engine components, brakes, suspension components
Aluminum	8.5	Engine block, wheel, radiator
Copper	1.5	Wiring, electrical components
Polymers and polymer matrix composites	9	Interior components, electrical and electronic components, under-the-hood components, fuel line components
Elastomers	4	Tires, trims, gaskets
Glass	3	Glazing
Other	10	Carpets, fluids, lubricants, etc.

Table 2.2 Use of steels (in weight %) in North American automobiles

Steel	Year	
	2007	2009
Plain carbon steels (mild steels)	57.8	52.6
High strength steels (HSS)	32.9	33.7
Advanced high strength steels (AHSS)	9.3	13.7

the curb weight in 1970 to nearly 8.8% in 2010 and is projected to reach 10% or higher in 2020. Much of this growth in the use of aluminum alloys has occurred at the expense of cast iron in engine and transmission components and copper-based alloys in radiators; but aluminum alloys, because of their lower density than steel's, are also making inroads in body panels and structures. The growth in the use of polymer matrix composites has also occurred due to their lower density. Among the polymer matrix composites, glass fiber composites are selected for most interior applications today, but they are also found in some exterior body panel or structural applications. Lighter components can be produced with carbon fiber composites, but because of their high cost, carbon fiber composites are not used in today's automobiles except in a few low-production volume, high-cost vehicles. With greater emphasis on vehicle weight reduction, it is expected that other lightweight materials, such as magnesium alloys, titanium alloys and carbon fiber composites will find several niche applications in future automobiles (Powers, 2000).

Table 2.3 lists the tensile properties of a few selected materials that are in competition with steels, and are either already being used in current vehicles or are likely to be used for future vehicle construction. At present, many of them are not cost-competitive with steels, and there are many technical and

Table 2.3 Material property comparisons

Material	Density (ρ) (g/cm^3)	Tensile modulus (E) (GPa)	Yield strength (S_y) (MPa)	Tensile strength (S_t) (MPa)	Coefficient of thermal expansion $(10^{-6}/°C)$
DQ low carbon steel	7.87	207	186	317	11
DP 400/700 steel	7.87	207	400	700	11
TRIP 450/800 steel	7.87	207	450	800	11
5182-H24 aluminum	2.7	70	235	310	23
6111-T62 aluminum	2.7	70	320	360	23
AZ91 magnesium	1.8	45	160	240	26
Ti-6Al-4V titanium	4.43	114	827	896	9
304 stainless steel	7.9	200	241	614	17
Nitronic 30 stainless steel	7.86	193	393	862	16
High strength CFRE[1] (unidirectional)	1.55	138	–	1550	–
High modulus CFRE (unidirectional)	1.63	215	–	1240	–0.9 (L) 27 (T)
GFRE[2] (unidirectional)	1.85	39	–	965	6 (L) 19 (T)
CFRE (quasi-isotropic)	1.55	45.5	–	579	0.9
Sheet molding compound (SMC-R50)	1.87	16	–	164	14.8

Note: L is the longitudinal direction and T is the transverse direction
(1): CFRE is carbon fiber reinforced epoxy, (2): GFRE is glass fiber reinforced epoxy.

cost challenges to be overcome with some of these materials, particularly for large production volumes. Nevertheless, their technical viability and weight saving potential have been demonstrated in many concept vehicles and prototype designs, and now, they are appearing, albeit in smaller quantities, in some production vehicles.

The greatest opportunity for weight reduction exists in the body and chassis components, which comprise 60% of a vehicle's weight. Many new materials and manufacturing processes have been developed in the last 20–25 years to lighten the weight of the body structure, body panels and suspension components. Powertrain weight, which includes both engine and transmission components, is between 25 and 30 percent of the vehicle weight. Several new materials and manufacturing process developments have been introduced to reduce the powertrain weight. The advanced materials considered for lightening the weight of major subsystems of a vehicle, such as body, chassis, suspension and powertrains are discussed in the following sections. Details of many of these materials can be found in a recently published book on materials, manufacturing and design for lightweight vehicles (Mallick, 2010).

2.2   Steels

Approximately 55% of the weight of US cars is made of steel. There are several advantages of using steel in auto body structures and body panels. The most important of them is its high modulus of elasticity, which at 207 GPa, is the highest among the structural materials considered for automotive applications. Steel is also the most inexpensive structural material available today. The wide variety of strengths available with steel, ranging from 200 MPa to 1500 MPa, is also an important advantage, since it gives an opportunity to select steel according to the structural design need. The use of high strength steels allows not only the downsizing of gage thickness, but improves the load carrying capacity and crashworthiness of the vehicle structure. Furthermore, steel's superior formability compared to aluminum and magnesium alloys, excellent weldability and recyclability are some of the reasons for steel's predominance in today's automobiles.

The automotive steel scenario has changed significantly in the last 25 years. Improvements in steel making processes (e.g., vacuum degassing and inclusion control) have made it possible to produce steel more cost effectively with much lower impurity levels (only about 10–20 ppm compared to 200–400 ppm by the traditional processes). Combination of new alloying techniques and improved thermo-mechanical processes, such as continuous annealing and controlled hot rolling, are now used to produce not only a broad spectrum of strength and ductility, but also better surface qualities and more uniform properties in sheet steels. Better corrosion resistance is achieved by new types of zinc alloy coatings (e.g., Zn-Fe and Zn-Ni) as well as new methods of applying them on the steel sheet surfaces (e.g., by electro-deposition instead of hot dipping). A relatively new process, called galvanneal, is able to produce superior corrosion resistance, formability as well as weldability of coated sheet steel. Laminated sheet steel with steel outer skins and a thin viscoelastic constrained layer (typically 0.025 mm thick) is available for noise and vibration control purposes (Yang *et al.*, 2001). One trade name for currently used laminated steel is Quiet Steel.

Sheet steels for body panels and body structures can be classified as plain carbon steels, high strength steels (HSSs) and advanced high strength steels (AHSSs) (Table 2.4). Among the plain carbon steels are the traditional mild steels, such as drawing quality (DQ) steels or drawing quality, aluminum-killed (DQAK) steels, and interstitial-free (IF) steels. Carbon-manganese (CMn) steels, bake-hardenable (BH) steels, solution-strengthened steels (SSS) and high strength low alloy (HSLA) steels belong to the second category. AHSSs, which have tensile strengths higher than 550 MPa, include dual phase (DP) steels, transformation-induced plasticity (TRIP) steels, complex phase (CP) steels, martensitic (MS) steels and hot-formed boron steels.

Among the high strength steels, BH steels have a good balance of yield

Table 2.4 Properties of several steels selected for body applications

Steel	Yield strength (MPa)	Tensile strength (MPa)	% Elongation	Strain hardening exponent (n)	Plastic strain ratio ($\bar{r}$)
DQ steel	186	317	42	0.22	1.5
BH 210/340	210	340	34–39	0.18	1.8
IF 300/420	300	420	29–36	0.20	1.6
HSLA 350/450	350	450	23–27	0.14	1.1
DP300/500	300	500	30–34	0.16	1.0
DP400/700	400	700	19–25	0.14	1.0
DP700/1000	700	1000	12–17	0.09	0.9
TRIP 450/800	450	800	26–32	0.24	0.9
Mart 950/1200	950	1200	5–7	0.07	0.9
Boron steel	1100	1500	5–7	–	–

strength, formability and dent resistance, and are selected for exterior body panels. BH steels are first cold-formed and then strengthened during paint baking, typically at 175 °C for 20 to 30 minutes. Depending on the strain hardening imposed by the cold forming operation prior to paint baking, increase in yield strength is in the range of 30–50 MPa, which occurs due to strain aging during the paint baking cycle. Another high strength steel is high strength low alloy (HSLA) steel, which achieves its high yield strength (300–550 MPa) due to the presence of fine-grained ferritic microstructure and small amounts (in the range of 0.005%) of carbide and nitride forming alloying elements, such as vanadium, niobium and titanium. The carbon content in HSLA steels is restricted to a maximum of 0.13% for improved formability and weldability. HSLA steels are selected for structural applications, such as cross beams and door intrusion beams.

DP steels contain martensitic dispersion of 20% to 70% by volume in a soft ferrite matrix. DP steels have lower yield strength than HSLA steels, but higher strain hardening capacity, which also gives them a higher tensile strength. The martensite content in DP steels, which determines their strength, can be varied by controlling the cooling rate during thermo-mechanical processing of these steels. The DP steels, on account of their high strength, are able to provide high energy absorption during crash events. A variation of the dual phase steels, containing bainitic dispersion instead of martensitic dispersion in a ferrite matrix, is also available. These are known as stretch flangeable steels, since they provide a higher resistance to edge cracking, and therefore, are more suitable for applications where flanged holes are needed. TRIP steels contain at least 5% by volume of retained austenite in addition to martensitic and bainitic dispersions in a soft ferrite matrix. The retained austenite in TRIP steels provides good formability. During the cold forming operation, the retained austenite transforms into martensite with increasing strain, thereby increasing its work hardening rate as well as strength. Unlike plain carbon steels and HSLA steels, both DP and TRIP steels have the potential of gaining higher strength during the paint baking cycle after cold forming.

Martensitic steels contain close to 100% martensite and small amounts of bainite and/or ferrite, and depending on the carbon content, exhibit tensile strengths ranging from 900 to 1,700 MPa. Because of high martensite content, they exhibit very low % elongation to failure, typically 5% or lower. These steels are selected for door beams and roof cross beams that are designed to prevent intrusion into the passenger compartment in case of side impact or rollover accidents. Hot-formed boron steels, which also have a martensitic structure and tensile strengths ranging up to 1,500 MPa, contain approximately 1.2% manganese and small amounts of boron (between 0.0005% and 0.001%). The martensite formation in these steels takes place during the hot forming operation which involves heating the steel in the pre-martensitic form to 930 °C and then transferring it to the forming die where 'in-situ' martensite formation takes place as the steel part is shaped and quenched to room temperature. The tensile strength of boron steels increases from approximately 400 MPa before hot-forming to 1,500 MPa after hot-forming.

Although AHSSs have much lower ductility and formability than conventional low carbon or high strength steels, they provide much higher crush resistance because of their high strength and are increasingly being selected for the front end structure, roof structure and other crash safety related structures of vehicles. As Fig. 2.1 shows, percentage elongation, which is

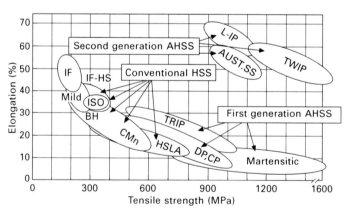

2.1 Ductility (represented by percentage elongation) vs tensile strength of various automotive sheet steels. (AUST.SS: austenitic stainless steel, BH: bake hardenable, C-Mn: carbon-manganese, CP: complex phase, DP: dual phase, HSLA: high strength low alloy, IF: interstitial free, IF-HS: interstitial free-high strength, ISO: isotropic, L-IP: lightweight steels with induced plasticity, MART: martensitic, Mild: mild steels, TRIP: transformation-induced plasticity, TWIP: twinning-induced plasticity).

often used as a measure of ductility, decreases with increasing strength of steels. General formability characteristics, such as strain hardening exponent (n) and plastic strain ratio (r), are lower for advanced high strength steels. AHSSs also have higher post-forming springback than lower strength steels, and unless proper part and tool design procedures are followed, may exhibit edge cracking at hole boundaries or near sheared edges.

There is another emerging category of steels, called second generation advanced high strength steels that have both high strength and ductility. One of the steels in this category is called twinning-induced plasticity (TWIP) steels (Cornette et al., 2005). TWIP steels have a high manganese content (typically 17% to 24%) that causes their microstructure to be fully austenitic at room temperature. During cold forming, deformation by twinning causes the austenite grains to break into finer sizes, thus increasing their strength by grain refinement. As shown in Table 2.5, TWIP steels have high strength as well as high percentage elongation, but are much more expensive than other steels.

Improvements have also taken place for the forging quality steels used for powertrain, suspension and steering components (Yamagata, 2005; Cho et al., 1994). One of these developments is microalloyed steels containing 0.3 to 0.6% C. The microalloying element is usually a small amount of vanadium (~ 0.05 to 0.15%), which forms vanadium carbide and nitride precipitates as the forged component is air cooled after hot forging. Tempering after air cooling is not necessary, since the precipitates in the relatively soft ferrite and pearlite matrix strengthen the steel. Microalloyed steels possess a good combination of strength and toughness, which can be further improved by grain size refinement through proper control of the inclusions in the steel as well as the forging conditions. The yield strength and percentage elongation of microalloyed steels are higher than conventional forging quality steels of similar carbon content. The fatigue strength is also higher. Furthermore, microalloyed steels do not need quenching and tempering, which not only reduces the cost, but also reduces the possibility of thermal distortion which

Table 2.5 Comparison of tensile properties of a TWIP steel with HSLA, DP and TRIP steels (in the transverse direction)

Steel	Yield strength (MPa)	Tensile strength (MPa)	Uniform elongation (%)	Total elongation (%)	Strain hardening exponent (n)
HSLA 320	342	428	16	24	0.169
DP 590	342	629	16	26	0.152
DP 980	644	1009	10	15	0.112
TRIP 800	478	825	21	27	0.216
TWIP 1000	496	1102	52	52	0.415

Source: Cornette et al. (2005)

results from quenching and tempering. The application of microalloyed forging steels includes connecting rods, wheel hubs, steering knuckles, etc.

2.3   Light alloys

2.3.1   Aluminum alloys

The most important advantage of aluminum alloys over steels in automotive applications is their low density, which is 2.7 g/cm^3 compared to 7.87 g/cm^3 for steels. Thus, the density of aluminum alloys is approximately 65% lower than that of steels. However, the modulus of aluminum alloys is 70 GPa compared to 207 GPa for steels, which means that for equal bending stiffness, an aluminum component will be 43.5% thicker than a steel component. As a result, the weight reduction achieved by aluminum will not be in the same proportion as the density ratio between the two materials. A simple weight calculation will show that substituting a steel body panel with an aluminum body panel will result in approximately 50% weight saving.

Both cast and wrought aluminum alloys are used in numerous applications in automobiles. The cast aluminum alloys are used mostly for engine, transmission and suspension components, whereas wrought aluminum alloys, in the form of sheet and extrusions, are used in body structure components and body panels. Another application area for aluminum alloys, such as AA 1200 and AA 3005, is in heat exchangers, which include radiator and condenser tubes and fins. The advantages of using aluminum in these applications is not only their high thermal conductivity, but their significantly higher strength-to-density ratio compared to that of copper-based alloys, which have been traditional materials for heat exchanger applications.

Cast aluminum alloys are mostly the 300-series (Al-Si-Cu or Al-Si-Mg) alloys, such as 319 for intake manifolds, cylinder heads and transmission housings, 383 for engine blocks, 356 for cylinder heads, and A356 for wheels and suspension arms. The principal alloying element in these alloys is silicon (Si), which contributes to their high fluidity. They can be cast using a variety of conventional casting techniques ranging from sand casting and die casting to more intricate permanent mold and lost foam/lost wax casting. Vacuum die casting, squeeze casting and semi-solid casting are used if higher casting integrity and fewer casting defects are desired. In addition to the 300-series alloys, a number of 200-series (Al-Cu) cast aluminum alloys, such as 201, 204 and 206, are used in chassis, suspension and engine components, such as brake calipers and connecting rods. Both 200- and 300-series alloys are heat treatable alloys.

The sheet aluminum alloys used for body panel and body structure applications are the work-hardenable 5000-series (Al-Mg) alloys, such as 5182, 5454 and 5754, and the age-hardenable 6000-series (Al-Mg-Si) alloys, such as 6009, 6022, 6061 and 6111 (Table 2.6). The 5000-series alloys are non-heat treatable, i.e., they cannot be strengthened by heat treatment. Sheets of these alloys are supplied in annealed O-temper condition and are strengthened by work hardening during the press forming operation. Sheets of 6000-series alloys are supplied in solution-annealed and naturally aged T4 condition and are usually strengthened to T6 condition by age hardening as they are being painted in the paint baking oven. The 5000-series alloys are highly formable, but since stretcher strain marks (Lüder's bands) may appear on the surface of these alloys during the forming operation, they are not selected for outer body panels. Instead, they are used in internal body panels and body structures. The 6000-series alloys, on the other hand, are resistant to stretcher strain marking and are used for both inner and outer body panels as well as body structure components.

The extruded automotive aluminum alloys are 6000-series (Al-Mg-Si) alloys, such as 6005, 6061, 6063, and 6082, and 7000-series (Al-Zn-Mg) alloys, such as 7004, 7116, 7029 and 7129. They are used in a variety of body structure and powertrain applications, such as cross members, front fender rails, engine cradles, seat frames, bumper beams and drive shafts. Both 6000- and 7000-series alloys are heat treatable by solution annealing followed by either natural aging or artificial aging. The 7000-series alloys are more difficult to extrude than the 6000-series alloys, especially in complex hollow shapes. They are also less corrosion resistant and weldable. In general, aluminum alloys can be extruded relatively easily compared to steel. Body structural parts, such as roof rails, require multiple stampings and welding when they are made out of steel. With aluminum, a single extruded aluminum section can be used. The use of a one-piece extruded section instead of a stamped and welded section can result in tooling and assembly cost reductions.

In general, formability of aluminum alloys is about two-thirds that of DQ steels. Due to lower formability, complex aluminum body panels may require several stamping steps or may have to be produced by assembling several

Table 2.6 Properties of several wrought aluminum alloys selected for body applications

Material	Yield strength (MPa)	Tensile strength (MPa)	% Elongation	Strain hardening exponent (n)	Plastic strain ratio ($\bar{r}$)
5182-O	130	275	24	0.33	0.80
5454-O	115	250	22	0.30	0.80
5754-O	100	220	26	0.30	0.80
6009-T4	125	220	25	0.22	0.64
6009-T62	260	300	11	–	–
6111-T4	150	280	26	0.28	0.70
6111-T62	320	360	11	–	–
6061-T6	275	310	12	–	–

separate stampings. In addition, because of aluminum's lower modulus, aluminum parts exhibit higher elastic springback after forming, and therefore, poorer shape retention than steel parts. Aluminum alloys have a greater tendency to gall than steel during the stamping operation and require larger amounts of lubrication as well as better die surface finish than steel.

Although aluminum alloys can be resistance spot welded like steel, higher welding currents are needed for aluminum alloys because of their low electrical resistivity and high thermal conductivity. The welding current for aluminum alloys is between 15 and 30 kA compared to 8–10 kA for steel. This means larger welding machines are needed for spot welding aluminum alloys and energy consumption is also higher. Fusion welding techniques, such as MIG welding, can also be applied to aluminum alloys, but due to their high thermal conductivity, high heat energy is needed. The two newly developed welding techniques that apply well to aluminum alloys are linear friction stir welding and friction stir spot welding. Other joining techniques that are being used with aluminum alloys are self-piercing riveting, clinching, adhesive bonding and weld-bonding (a combination of spot welding and adhesive bonding).

2.3.2   Magnesium alloys

Magnesium alloys are considered for automotive applications principally because of their very low density (1.74 g/cm^3 compared to 2.7 g/cm^3 for aluminum alloys). They also have a higher strength-to-weight ratio compared to aluminum alloys. On the other hand, the modulus of magnesium alloys is 45 GPa, which is significantly lower than that of steel and aluminum alloys; however, because of their low density, the modulus-to-density ratio of magnesium alloys is the same as that of aluminum alloys. Magnesium alloys have low ductility and poor formability; but many magnesium alloys can be cast in thin sections that are as low as 2 mm in thickness. The common manufacturing method for making magnesium automotive components is die casting, which allows the opportunity for parts consolidation and cost reduction.

Like aluminum alloys, magnesium alloys can be divided into casting alloys and wrought alloys. Among the casting alloys, AZ91, with aluminum and zinc as the principal alloying elements, is used in many non-structural components, such as brackets, covers and housings where it provides significant weight saving over aluminum alloy A380. For structural components where higher ductility and crash resistance are important, such as instrument panel beams, steering wheel armatures and seat structures, AM20, AM50 or AM60 are used. The principal alloying elements in the AM-series alloys are aluminum and manganese. Among the wrought magnesium alloys, AZ80 is used for extruded sections and AZ31 is used for sheets. The yield strengths of these alloys are comparable to the yield strengths of 5000- and 6000-series aluminum alloys, but they are less ductile than aluminum alloys. The room temperature formability of wrought magnesium alloys is also much lower than that of aluminum alloys and steel. Because of this, elevated temperatures, in the range of 200–400 °C, are recommended for sheet stamping, bending and other forming operations with AZ31 (Luo, 2005). Elevated temperatures are also used for the extrusion of AZ80.

One major concern with magnesium alloys is their poor corrosion resistance. While the aqueous corrosion resistance of AZ and AM alloys in a salt environment is comparable to that of cast aluminum alloys, their galvanic corrosion resistance is very poor. Thus, when a magnesium component is attached to a steel component or when two magnesium components are joined together using a steel fastener, magnesium is aggressively corroded. Since the galvanic corrosion of magnesium in the presence of aluminum is much less, aluminum washers, aluminum fasteners or aluminum-coated steel fasteners are often used with magnesium. In the die cast AM50 radiator support assembly in Ford F150 light trucks, the galvanic corrosion protection from the attached steel brackets was achieved by a combination of surface coatings and 0.7 mm thick aluminum (AA5052) isolators placed between the magnesium and steel components (Balzer et al., 2003).

Magnesium alloys are currently being considered for several powertrain applications, such as transmission cases and engine blocks. AZ91 alloy is selected for manual transmission cases where the operating temperature is below 120 °C. The operating temperature of automatic transmission cases and engine blocks can reach up to 200 °C. AZ91 or other conventional casting alloys are not suitable for these applications, since they exhibit significant creep at temperatures higher than 125 °C. Due to creep, the clamping load in the bolted joints of these alloys is reduced, which may cause gas and oil leaks and also increase noise and vibration. Recently, several creep-resistant magnesium alloys containing rare earth elements and alkaline earth elements have been developed which show promising bolt load retention and are considered better suited for powertrain applications (Pekguleryuz and Kaya, 2003). Some of these creep-resistant alloys are considered good candidates for engine block, oil pan and other engine components, and are being considered in developing a magnesium-intensive engine with a potential weight saving of at least 15% over a conventional aluminum-intensive engine (Hines et al., 2006).

2.3.3   Titanium alloys

The principal advantages of titanium alloys are their low density, high strength-to-density ratio, excellent corrosion resistance and superior strength retention at elevated temperatures ranging up to 500 °C. The density of titanium is 4.43

g/cm^3, which is higher than that of aluminum alloys, but significantly lower than that of steel. The modulus of titanium is 114 GPa, which is also higher than that of aluminum alloys, but nearly half the modulus of steel.

The major drawback of titanium for automotive applications is its high cost compared to steel, aluminum and magnesium. On a unit weight basis, the cost of sheet titanium is \$18 to \$110 per kg compared to only \$0.70 to \$1.30 per kg for sheet steel and \$2.20 to \$11 per kg for sheet aluminum (Froes *et al.*, 2004). On the basis of cost, titanium is not expected to compete with steel or aluminum in body panel or body structure applications.

However, the potential for saving weight using titanium exists in several automotive applications. One of these applications is the suspension coil springs, where titanium's relatively low shear modulus and excellent fatigue strength give it an advantage over steel. Since spring deflection is inversely proportional to the shear modulus, a titanium coil spring can be designed with fewer active coils than a steel coil spring, which contributes not only to weight reduction, but also to increasing natural frequency of vibration. Titanium coil springs have been used in aircraft for many years. The first titanium coil spring in the automotive industry appeared in 2001 in the Volkswagen Lupo FSI (Faller and Froes, 2001). The titanium alloy for the VW springs was a Ti-4.5 Fe-6.8 Mo-1.5 Al alloy (Timetal LCB), which is 50% lower in cost than conventional α/β and β-titanium alloys and was specifically developed for automotive applications. The titanium coil springs were about 60% lighter than the steel coil springs they replaced.

Titanium's high strength-to-density ratio, fatigue strength and strength retention at elevated temperatures can be utilized to reduce the weight of reciprocating engine components, such as connecting rods, pistons and piston pins. Other engine components where titanium has performed well are engine valves, valve retainers and valve springs. The reduced mass of many of these engine components has the secondary effect of reducing friction, which in turn improves engine efficiency. For example, it is estimated that the use of a titanium valve system can reduce the engine frictional loss by about 10%, which, for a typical driving cycle amounts to 3–4% improvement in fuel economy (Sherman and Allison, 1986). Titanium matrix composites and titanium aluminide, which is an intermetallic compound of titanium and aluminum, are selected for engine valve components, which undergo a significant amount of wear.

Another potential application area of titanium is in the exhaust system, since titanium possesses excellent oxidation resistance up to 700 °C. Due to its lower density, considerable weight saving can be achieved over stainless steel which is currently used for tail pipe, muffler and other components in the exhaust system. Titanium mufflers, offered as an option in the Corvette Z06, were 41% lighter than stainless steel mufflers. Since many of the exhaust system components are cold formed, unalloyed (commercially pure) titanium (grade 1 or 2) is recommended, since it has better strain-to-failure and formability than the α/β or β titanium alloys. However, unalloyed titanium is more suitable for the rear section of the exhaust system, where the temperature is considerably lower than that of the front section.

2.4   Stainless steels

The density and modulus of stainless steel are very close to the density and modulus of steel, and therefore, in stiffness-critical applications, direct substitution of steel with stainless steel does not produce any weight reduction. In strength-critical applications, stainless steel can provide weight reduction over steel for the following reasons.

- The yield strength-to-density ratio of several stainless steels is higher than that of high strength steels.
- Stainless steel has a higher work hardening coefficient and formability than steel, which means it can tolerate higher uniform plastic deformation and thickness reduction during forming.
- Stainless steel has a higher strain rate sensitivity than steel, which means it can absorb higher crash energy than steel. Additionally, it also has the capability of collapsing progressively in a controlled manner.

Another great advantage of stainless steel is its corrosion resistance. Anti-corrosion coatings may not be needed if stainless steel is used instead of steel. Despite the above advantages, stainless steel has found very little application in automotive structure because of its high cost. A few structural applications where stainless steel has been tried are fuel tanks, knuckle arms and wheels.

Stainless steel is available in a variety of grades, but the two grades that are used for automotive applications are the austenitic grade (300 series alloys, containing Cr and Ni as the principal alloying elements) and the ferritic grade (400 series alloys, containing Cr as the principal alloying element). The austenitic grade is non-magnetic and has higher yield strength, ductility and corrosion resistance then the ferritic grade. Neither grade can be strengthened by heat treatment, but both grades can be strengthened by cold work. The austenitic grade has a higher formability than the ferritic grade. A nitrogen-strengthened version of the austenitic grade, called nitronic, is also available, and several nitronic alloys (e.g., Nitronic 19D and Nitronic 30) can be used for automotive applications. Nironic 19D is a casting alloy and is recommended for suspension components. Nitronic 30 has excellent formability and is recommended for body panels.

The principal use of stainless steel in today's automobiles is in the exhaust system, where its exceptional corrosion and oxidation resistances give it a considerable edge over steel or aluminized steel. Typical choices for the hot end of the exhaust system, which includes the exhaust manifold, down pipe and catalytic converter, are the austenitic grades, such as 309 or 310 (25% Cr, 20% Ni). For the cold end, which includes the resonator, intermediate pipe, silencer and tail pipe, either austenitic grades, such as 304 (18% Cr, 9% Ni) or ferritic grades, such as 409 (12% Cr), are selected.

2.5   Cast iron

With increasing use of high strength steels and light non-ferrous alloys, the cast iron content in automobiles has decreased considerably over the last few years. Cast iron, due to its density as high as that of steel, does not offer any weight saving advantage. Furthermore, cast iron is a low ductility material. The principal advantages of cast iron for which it continues to be used are its low cost, high wear resistance, damping and excellent machinability.

Cast iron is used in many engine applications. One of these engine applications is the cylinder block. Although aluminum is increasingly used for making cylinder blocks in gasoline engines, grey cast iron is still the predominant material for cylinder blocks in diesel engines. With increasing trend toward smaller engines and higher in-cylinder pressures, compacted graphite iron (CGI) is finding increasing use instead of grey cast iron. The graphite particles in CGI are in vermicular or worm-like form instead of the flaky form observed in grey cast iron or spherical form observed in nodular cast iron. As a result, the properties of CGI fall between grey cast iron and nodular cast iron. The tensile strength of CGI is 1.5 to 2 times higher than that of grey cast iron, and, the modulus of CGI is 150 GPa compared to 105 GPa for grey cast iron. The thermal conductivity of CGI is lower: 38 W/m-°K compared to 48 W/m-°K for grey cast iron. With higher strength, higher modulus and lower thermal conductivity, CGI cylinder blocks can be designed with lower thickness than grey cast iron cylinder blocks.

Cast irons used in structural automotive applications are ductile irons, which have high yield strength (275–625 MPa) and relatively high ductility (2–18% elongation). The modulus of ductile irons is between 160–170 GPa, which is considerably higher than that of aluminum. Ductile irons are used in steering knuckles, brake calipers, crank shafts, cam shafts and many other powertrain components. Austempered ductile iron (ADI), produced by a heat treatment process called austempering, has a significantly higher yield strength (400–1200 MPa) and higher fracture toughness than conventional ductile irons. The yield strength-to-density ratio of ADI is considerably higher than that of cast or forged aluminum. This is the reason for selecting ADI over aluminum alloys in many chassis and suspension components.

2.6   Composite materials

2.6.1   Polymer matrix composites

Polymer matrix composites (PMCs) are prepared by combining high-strength, high-modulus fibers, such as glass, carbon and Kevlar fibers, with either a thermoplastic or a thermoset polymer matrix. Depending on the design requirements and the properties desired, fibers can be used in a variety of lengths (continuous or discontinuous) and orientations (Mallick, 2008). With unidirectional continuous fibers (i.e., all fibers are oriented in the same direction), the modulus and strength of the composite are highest in the fiber direction (longitudinal direction), but lowest normal to the fiber direction (transverse direction). For example, the longitudinal modulus of a unidirectional high modulus carbon fiber reinforced epoxy is 207 GPa (which is equivalent to the modulus of steel), whereas the transverse modulus is only 14 GPa. Bi-directional reinforcement (e.g., fabric reinforcement) produces a more balanced set of strength and modulus in the two fiber directions (called warp and weft directions, which are 90° to each other); however, they are lower than the longitudinal strength and longitudinal modulus of a unidirectional composite. If the fibers are randomly oriented, the properties are the same in all directions in the plane of the composite; however, they are significantly lower than the properties of composites containing either unidirectional or bidirectional continuous fibers. Thus, unidirectional and bi-directional composites behave as non-isotropic materials, whereas random fiber composites behave as an isotropic material.

Most of the polymer matrix composites in today's automobiles contain randomly oriented discontinuous glass fibers. They are manufactured using either injection molding or compression molding processes. E-glass fibers are selected because of their much lower cost than carbon or Kevlar fibers. Because of the discontinuous fiber lengths and random fiber orientation, they do not provide the highest strengths and modulus that can be achieved with continuous fiber composites. Continuous fiber composites, in general, have higher strength-to-density ratio and higher modulus-to-density ratio than steel and light non-ferrous alloys. They also have excellent fatigue strength and fatigue damage tolerance. The possibility of making laminated structures with different fiber orientations in different layers of the laminate or making a sandwich structure with high modulus composite skins and low density foam, balsa wood or aluminum honeycomb in the core provides a tremendous design flexibility that does not exist with metals.

The automotive applications of PMC include both thermoplastic matrix composites and thermoset matrix composites. Thermoplastic matrix composites are used for a variety of interior and body applications, such as instrument panels, seat backs, inner door panels, fender aprons and bumper beams. The thermoplastic polymers in these applications are usually polypropylene (PP), polybutylene terephthalate (PBT), polycarbonate/ABS blends, polyamide-6 or polyamide-6,6. They are selected because of their relatively low cost compared to high performance thermoplastics, such as polysulfone, poly ether ether ketone (PEEK), etc.

Most of the thermoplastic matrix composites used today can be classified as short fiber composites (SFT). The fiber length in these composites is in the order of 1 mm. Recent developments of long fiber thermoplastics (LFT), glass mat thermoplastics (GMT) and commingled fabric reinforced thermoplastics have increased the possibility of using thermoplastic matrix composites in several structural applications, such as interior door panels, bumper beams and cross members. Technology of making thermoplastic matrix composite laminates with continuous fibers is also evolving, which will make it possible to use them in structural applications. Another type of thermoplastic matrix composites that have appeared in the market in recent years is called self-reinforced thermoplastics (SRTs). They are single polymer composites in which the materials for the reinforcing fibers and the matrix are of the same thermoplastic polymer type; for example, polypropylene fibers in polypropylene matrix. Thus, unlike glass or carbon fiber reinforced thermoplastics, SRTs are completely recyclable. Since the fibers and the matrix are of the same chemical structure, a strong interfacial bond exists between the two, which helps in achieving high tensile strength for the composite. The density of self-reinforced polypropylene is significantly lower than that of glass fiber reinforced polypropylene (Table 2.7). One outstanding property of self-reinforced polypropylene is its high impact strength, which is nearly three times higher than that of continuous glass mat reinforced

Table 2.7 Properties of several polypropylene matrix composites

Properties	Self-reinforced thermoplastics	Glass mat thermoplastics (GMT)		Long fiber thermoplastics
	Bi-directional polypropylene fabric	Randomly oriented continuous fibers (with 40 wt.% E-glass fibers)	Unidirectional fibers (42 wt.% E-glass fibers)	Randomly oriented long fibers (with 40 wt.% E-glass fibers)
Tensile strength (MPa)	207	108	276	117
Modulus (GPa)	6.4	5.82	10.1	8.96
Strain at failure (%)	5.7	2.5	2.5	2–3
Density (g/cm^3)	0.78	1.21	1.24	1.21

polypropylene. Self-reinforced polypropylene is currently being investigated for seat frames and door panels.

The most common thermoset matrix composite used in the automotive industry is sheet molding compound (SMC), which contains randomly oriented discontinuous E-glass fibers (typically 25 mm long) in a thermoset polymer, such as a polyester or a vinyl ester resin. Examples of SMC parts are hoods, deck lids, fenders, radiator supports, bumper beams, roof frames, door frames, engine valve covers, timing chain covers, oil pans, etc. These parts are produced by the compression molding process. Another manufacturing process used for making thermoset matrix composite parts is called structural reaction injection molding (SRIM). The matrix in composites produced by SRIM is either polyurethane or polyurea.

SMC usage has experienced a large growth in the automotive industry over the last 25 years. Its advantages over steel include not only the weight reduction, but also lower tooling cost and parts consolidation. The tooling cost for compression molding SMC parts is 40–60% lower than that for stamping steel parts. An example of parts consolidation can be found in radiator supports in which SMC is used as a substitution for low carbon steel. The composite radiator support will typically be made of two SMC parts bonded together by an adhesive instead of 20 or more stamped steel parts assembled together by a large number of screws. Another example of parts consolidation can be found in the station wagon tailgate assembly, which has significant load-bearing requirements in the open position. The composite tailgate consists of two pieces, an outer SMC shell and an inner reinforcing SMC piece. They are bonded together using a urethane adhesive. In one such application, the SMC tailgate replaced a seven-piece steel tailgate assembly, at about one-third its weight.

Among the chassis components, the first major structural application of polymer matrix composites is the Corvette rear leaf spring, introduced first in 1981 (Kirkham *et al.*, 1982). A uni-leaf E-glass fiber reinforced epoxy spring was used with as much as 80% weight reduction as compared to a multi-leaf steel spring. Other structural chassis components, such as drive shafts and road wheels, have been successfully tested in laboratories and proving grounds. They have also been used in limited quantities in production vehicles. They offer opportunities for substantial weight savings, but so far they have not proven to be cost-effective over their steel or aluminum counterparts.

While glass fibers are the primary reinforcing fibers used in today's automotive composites, it is well recognized that much higher weight reduction can be achieved only if carbon fibers are used. Carbon fiber reinforced polymers have much higher modulus-to-density and strength-to-density ratios than glass fiber reinforced polymers (Table 2.8). The reason for not using carbon fibers in today's vehicles is that the current carbon fiber price, at $16/

Table 2.8 Material indices and relative costs

Material	Properties (relative values)			Material index for stiffness critical design			Material index for strength critical design		Cost of material (relative values)
	Density (ρ)	Modulus (E)	Tensile strength (S_t)	Tension $\dfrac{E}{\rho}$	Buckling $\dfrac{E^{1/2}}{\rho}$	Bending $\dfrac{E^{1/3}}{\rho}$	Tension $\dfrac{S_t}{\rho}$	Bending $\dfrac{S_t^{1/2}}{\rho}$	
DQ Steel	1	1	1	26.3	1.83	0.75	40.3	2.26	1
DP Steel	1	1	2.21	26.3	1.83	0.75	88.9	3.36	1.15
AA6111	0.34	0.34	1.13	25.9	3.10	1.53	133.3	7.03	4–5
Mg AZ91	0.23	0.22	0.76	25	3.73	1.97	133.3	8.6	4–5
SS304	1	0.97	1.94	25.3	1.79	0.74	77.7	3.14	6–8
High strength CFRE	0.2	0.67	4.89	89	7.58	3.33	1000	25.4	15–20
GFRE	0.23	0.19	3.04	21.1	3.37	1.83	521.6	16.8	8
SMC	0.24	0.08	0.52	8.5	2.14	1.35	87.7	6.85	1.5

kg or higher, is not considered cost-effective for automotive applications. Many development projects in the past have demonstrated the weight saving potential of carbon fiber reinforced polymers; unfortunately, most of these projects did not go beyond the prototyping and structural testing stages due to the high cost of carbon fibers and the lack of manufacturing processes suitable for mass production of composite parts. Recently, several high priced vehicles have started using carbon fiber reinforced polymers in a few selected components. One recent example of this is seen in the BMW M6 roof panel, which was produced by a process called resin transfer molding (RTM). The material is a carbon fiber reinforced epoxy. This panel is twice as thick as a comparable steel panel, but is still 5.5 kg lighter. One added benefit of reducing the weight of the roof panel is that it lowers the center of gravity of the vehicle, which is an important design consideration for vehicle stability.

Carbon fiber reinforced polymers are used extensively in motor sports where a lightweight structure is essential for gaining the competitive advantage of higher speed (O'Rourke, 2000) and cost is not a major material selection decision factor. The first major application of these composites in race cars started in the 1950s when glass fiber reinforced polyester was introduced as replacement for aluminum body panels. Today, all major body, chassis, interior and suspension components in Formula 1 race cars utilize carbon fiber reinforced epoxy. One major application of carbon fiber reinforced epoxy in these cars is the survival cell, which protects the driver in the event of a crash. The nose cone located in front of the survival cell is also made of carbon fiber reinforced epoxy. Its controlled crush behavior is critical to the survival of the driver.

The major barrier to the application of carbon fiber reinforced polymers is the high material cost, which is solely due to the high cost of carbon fibers. It has been suggested that if carbon fiber cost reduces to $8–$10/kg, carbon fiber reinforced polymers will become a more viable material option for large-scale automotive applications. The largest contributors to the high cost of carbon fibers are the starting material or precursor cost and the cost of the energy-intensive thermal pyrolysis process used for making carbon fibers. Another current problem with carbon fibers is the availability. Much of the world's production of carbon fibers is consumed by the aerospace and sporting goods industries. New technologies are being developed to produce low-cost carbon fibers and to scale-up the production rate that can perhaps meet the automotive industry's need (Warren *et al.*, 2002).

Widespread use of polymer matrix composites, including carbon fiber reinforced polymers, will require the development of processing methods with a production cycle time that is competitive with that for steel. The cycle time for the molding processes used today for manufacturing structural automotive composite parts is between 1 and 5 minutes, compared to less than 10 seconds for stamping steel parts. Although the possibility of parts consolidation in composites may reduce the tooling and assembly costs, the processing cost due to higher cycle time causes the total manufacturing cost to be high. For building up confidence in polymer matrix composites for structural automotive applications, long-term durability data, reliable joining techniques, appropriate CAE design tools, and fast non-destructive inspection methods are also needed.

2.6.2   Metal matrix composites

Metal matrix composites (MMCs), by virtue of their low density, high strength-to-weight ratio, high temperature strength retention, and excellent creep, fatigue and wear resistances, have the potential for replacing cast iron and other materials in engines and brakes. Typically, MMCs considered for automotive applications contain either silicon carbide (SiC), aluminum oxide (Al_2O_3) or other ceramic particles or short fibers in a light alloy, such as aluminum, magnesium and titanium. MMCs have been developed for use in diesel engine pistons, cylinder liners, brake drums and brake rotors (Chawla and Chawla, 2006). Other potential applications where MMCs have been tried are connecting rods, piston pins and drive shafts. The major impediment toward their wider use is their high cost.

2.6.3   Nanocomposites

Nanocomposites contain nanometer (10^{-9}m) size reinforcements, such as nanoclay, carbon nanofibers and carbon nanotubes in a polymer matrix. The properties of these nano-reinforcements are considerably higher than conventional reinforcing fibers, such as glass and carbon fibers. Furthermore, their surface area to volume ratio is very high, which provides a greater interfacial interaction with the matrix. These composites show not only high modulus and strength, but also excellent thermal, electrical, optical and other properties, and in general, at relatively low reinforcement content.

Nanoclay is a platelet-type smectite clay mineral containing several layers of silicates. Each silicate layer is 1 nm thick and has a surface area of 100 nm^2 or higher. The most common smectite used in nanocomposites is called montmorillonite. To be an effective reinforcement, the silicate layers have to be exfoliated so that they are completely separated from each other and uniformly dispersed in the polymer matrix. The clay particles are chemically treated to promote the exfoliated dispersion.

A variety of techniques are available to mix nanoclay particles with a thermoplastic polymer. Melt mixing in an extruder or an injection molding machine is one of these techniques. The ability of montmorillonite to significantly improve modulus and strength of polyamide-6 was first reported by Toyota in 1987 (Okada and Usuki, 2007). The composite was prepared by *in-situ* polymerization. With the addition of only 4.2 wt.% of exfoliated montmorillonite, the tensile modulus of polyamide-6 was nearly doubled and its tensile strength increased by more than 50%. The heat deflection temperature was increased by 80 °C compared to polyamide-6. The first automotive application of this material was the timing belt cover in a Toyota car. Since then, nanoclay reinforced thermoplastics have found applications in engine covers, body side moldings, cargo floors and seat backs (Wang and Xiao, 2008). Polyamide-6 reinforced with only 2 wt.% nanoclay has five times the resistance to gasoline permeation compared to polyamide-6, which has prompted the use of this material in fuel lines.

Apart from nanoclay, a considerable amount of research is currently being conducted in developing carbon nanofiber as well as carbon nanotube reinforced polymers. Both types of reinforcement significantly increase the modulus and strength and decrease the coefficient of thermal expansion of the polymer. The other major benefit is the increase in electrical conductivity (Harris, 2004), which helps in dissipating static electricity build-up in electronic components and fuel lines and during on-line painting of thermoplastic body panels. At present, the use of these nano-reinforcements is relatively few, mainly because of their high cost and low availability.

2.7   Glazing materials

The glazing materials in a vehicle are laminated glass used for the windshield, and tempered glass used for side windows, rear window and sunroof. Laminated glass is constructed of two 1.8–2.3 mm thick sheets of glass with a very thin layer (typically 0.76 mm thick) of polyvinyl butyrate (PVB) in between. The PVB layer makes the windshield shatter-proof, which is essential for the safety of the driver and the front passengers. Tempered glass is a single sheet of glass (typically 2.4 to 2.6 mm thick) and is strengthened by heating it above the annealing point of 720 °C followed by rapid cooling. Tempered glass is much easier to penetrate than laminated glass and fractures in a brittle manner when impacted, but it is 3 to 4 times cheaper than laminated glass.

Although the weight of the glazing material is only 2–3 percent of the total weight of a vehicle, several alternatives are being considered to reduce its weight. One of these alternatives is to reduce the windshield thickness by using thinner glass sheets; however, a large reduction in the thickness may not only raise concern about safety, but also reduce its contribution to the torsional stiffness of the vehicle (which is approximately 10 percent with the current windshield thickness). Another alternative is to use polycarbonate instead of glass (Mori and Koursova, 2000), which has a density of 1.2 g/cm^3 compared to 2.5 g/cm^3 for glass. Polycarbonate is a transparent thermoplastic with optical properties comparable to glass. It is also a ductile polymer with high impact resistance. However, polycarbonate windshields require a scratch resistant coating on the surface. Since polycarbonate has a lower modulus than glass, polycarbonate windshields are thicker than glass windshields. They are also more expensive than glass windshields. Another possible material for glazing applications is laminated polymethyl methacrylate, which is being used in the side and front windows of a lightweight demonstrator Lotus Exige. It contains a soft inside layer between two sheets of polymethyl methacrylate and weighs half as much as glass windows.

2.8   Conclusions

This chapter has given a broad overview of advanced materials being considered for lightweight automotive structures and components. No one material has all the attributes to build lightweight automobiles needed for significant fuel economy improvement that will also meet stringent safety regulations, be environmentally friendly and remain cost effective. Therefore, it is expected that future automobiles will use a mix of materials that will include advanced high strength steels, aluminum and magnesium alloys, and carbon fiber composites. If that is how the future automobiles are going to be built, there are several design and manufacturing issues that need to be addressed. They include joining and assembly, corrosion and other interactions between dissimilar materials, recycling and life cycle values. Cost is another important factor that needs to be considered. Most of the advanced materials are more expensive than plain carbon steels and will require investment in dies and tools that are different from the ones used with plain carbon steels. Thus, while many studies have shown that prototype automobiles can be built with 30% to 50% weight saving compared to today's automobiles, it will require significant research and development as well as strong impetus before such weight saving will be realized in high volume, production automobiles.

2.9   References

Balzer J S, Dellock P K, Maj M H, Cole G S, Reed D, Davis T, Lawson T and Simonds G (2003), 'Structural magnesium front end support assembly', 2003 SAE World Congress, Paper no. 2003-01-0186, Warrendale, Society of Automotive Engineers.

Cho W-S, Kim K-S, Jo E-K and Oh S-T (1994), 'Development of medium carbon microalloyed steel forgings for automotive components', 1994 SAE World Congress, Paper no. 940784, Warrendale, Society of Automotive Engineers.

Chawla N and Chawla K K (2006), 'Metal-matrix composites in ground transportation', *Journal of Metals*, November, 67–70.

Cornette D, Cugy A, Hildenbrand A, Bouzekri M and Lovato G (2005), 'Ultra high strength FeMn TWIP steels for automotive safety parts', 2005 SAE World Congress, Paper no. 2005-01-1327, Warrendale, Society of Automotive Engineers.

Faller K and Froes F H (2001), 'The use of titanium in family automobiles: current trends', *Journal of Metals*, 27–28.

Froes F H, Friedrich H, Kiese J and Bergoint D (2004), 'Titanium in the family automobile; the cost challenge', *Journal of Metals*, 40–44.

Harris P J F (2004), 'Carbon nanotube composites', *International Materials Reviews*, **49**(1), 31–42.

Hines J A, McCune R C, Allison J E, Powell B R, Ouimet L J, Miller W L, Beals R, Kopka L and Reid P P (2006), 'The USAMP magnesium powertrain cast components projects', 2006 SAE World Congress, Paper no. 2006-01-0522, Warrendale, Society of Automotive Engineers.

Kirkham B E, Sullivan L S and Bauerie R E (1982), 'Development of the Liteflex suspension leaf spring', 1982 SAE World Congress, Paper No. 820160, Warrendale, Society of Automotive Engineers.

Luo A A (2005), 'Wrought magnesium alloys and manufacturing processes for automotive applications', 2005 SAE World Congress, Paper no. 2005-01-0734, Warrendale, Society of Automotive Engineers.

Mallick P K (2008), *Fiber Reinforced Composites*, Boca Raton, CRC Press.

Mallick, P K (ed.) (2010), *Materials, Manufacturing and Design for Lightweight Vehicles*, Cambridge, Woodhead Publishing.

Mori H and Koursova L (2000), 'Automotive glazing: issues and trends', 2000 SAE World Congress, Paper no. 2000-01-2691, Warrendale, Society of Automotive Engineers.

Okada A and Usuki A (2007), 'Twenty-year review of polymer-clay nanocomposites at Toyota Central R & D Labs'. 2007 SAE World Congress, Paper No. 2007-01-1017, Warrendale, Society of Automotive Engineers.

O'Rourke B P (2000), 'Formula 1 applications of composite materials', *Comprehensive Composite Materials*, **6**, 381–393.

Pekguleryuz M O and Kaya A A (2003), 'Creep resistant magnesium alloys for powertrain applications', *Advanced Engineering Materials*, 5, 12, 866–878.

Powers W F (2000), 'Automotive materials in the 21st century', *Advanced Materials and Processes*, **158**(5), 38–41.

Sherman A M and Allison J E (1986), 'Potential for automotive applications of titanium alloys', 1986 SAE World Congress, Paper no. 860608, Warrendale, Society of Automotive Engineers.

Wang Z and Xiao H (2008), 'Nanocomposites: recent development and potential automotive applications', 2008 SAE World Congress, Paper no. 2008-01-1263, Warrendale, Society of Automotive Engineers.

Warren C D, Shaffer J T, Paulauskas F L and Abdullah M G (2002), 'Low cost carbon fiber for the next generation of vehicles: novel technologies', 2002 SAE World Congress, Paper no. 2002-01-1906, Warrendale, Society of Automotive Engineers.

Yamagata, Y (2005), *The Science and Technology of Materials in Automotive Engines*, Cambridge, Woodhead Publishing.

Yang B, Nunez S W, Welch T E and Schwaegler J R (2001), 'Laminate dash Ford Taurus noise and vibration performance', 2001 SAE World Congress, Paper No. 2001-01-1535, Warrendale, Society of Automotive Engineers.

3

Advanced metal-forming technologies for automotive applications

N.J. DEN UIJL and L.J. CARLESS, Tata Steel RD&T,
The Netherlands

Abstract: In this chapter an overview is given of the forming technologies available to produce body and chassis parts in automotive manufacturing. First, a review is given of the metallurgical background of forming technology. Next the different forming techniques are presented and the materials available are discussed. Then some aspects of the modelling technology that has helped to advance forming technology in recent decades are discussed. The chapter closes with some economic considerations on the application of forming and materials for the automotive industry. Obviously a short chapter like this can never give full details on a subject so wide and essential to automotive manufacturing as forming technology, it could easily be expanded to a full volume, detailing various aspects, but it should give some insight about the subject and enable the reader to find information for further study.

Key words: sheet metal formability, sheet metal forming technology, numerical material data, finite element modelling of forming, economics of automotive forming operations.

3.1 Formability

After casting, materials are rolled into sheets. Although the basic chemical composition of a material is not changed after casting (except for some surface treatments that may follow) the mechanical characteristics determining formability of the material are primarily dependent on the thermomechanical treatment the material experiences after casting. Based on the chemical composition, the material may be hotrolled, coldrolled, heat treated, coated and coiled to finally achieve its desired properties. Not all materials will undergo all available treatments, but whatever treatment will be applied, it will have an effect on its formability. Therefore it is never sufficient to describe a sheet material by its chemical composition alone. Steel is never just steel and aluminium is never just aluminium. There is a combined drive for increased strength levels and decreased thicknesses in the automotive industry to comply with regulations to increase safety and decrease fuel consumption (e.g., through decreased weight). This has led to the successful introduction of new materials that depend on these thermomechanical treatments for their improved properties.

28

3.1.1   Stress and strain

Tensile tests are performed using a tensile test machine. A predefined test piece is placed in the grips and clamped. The grips are pulled apart. The rate can either be load controlled or displacement controlled. During the test the applied force is measured (using a load cell) as well as the displacement. Either the displacement of the grips or the displacement between two points on the test piece can be measured using an extensometer. Measuring the displacement of the grips may be inaccurate due to slip between the grips and the test piece.

The tensile curve starts with elastic deformation. The point where elastic deformation ends and plastic deformation begins is called the yield point. The corresponding stress is called the yield stress. Plastic strains are determined by fitting a line through the elastic portion of the curve. If the plastic strain is determined taking elastic strain into account, the parameter is designated with R_p. A general parameter is the strain at 0.2% plastic deformation, $R_{p0.2}$. If elastic deformation is not taken into account, R_t is used to designate the total strain (see Fig. 3.1).

Some materials show an elongated yield point, where a plateau is shown before further hardening occurs. This elongation can be characterised by a peak followed by a plateau. R_{eH} designates the upper yield stress; the maximum value of stress before a decrease in stress. R_{eL} designates the lower yield stress; the lowest value of stress during plastic yielding (ignoring transient effects). The yield point elongation can occur at a slope of slow increase in stress levels.

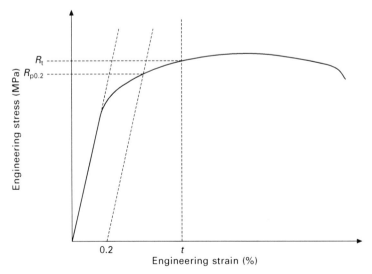

3.1 Schematic depiction of tensile curve.

The maximum stress a material can withstand is called ultimate tensile stress, often designated as R_m. This is not the maximum at the true stress–true strain curve, as further elongation will lead to necking. The plastic strain associated with R_m is A_g. If no correction is made for elastic deformation, A_{gt} is used (see Fig. 3.2).

The fracture strain is determined from the length of the specimen at breaking including the necked region. The fracture strain is dependent on the original gauge length of the test specimen, and this is indicated by a subscript to the maximum strain level; e.g., A_{80}.

Scatter in the measurement data can obscure the mechanical data. Therefore data is fitted to a (second order) polynomial (see Fig. 3.3). For calculation and simulation it is better to use these corrected data to avoid numerical irregularities.

True stress and strain

The stresses derived from the tensile curve are engineering stresses, calculated using the original cross-section area of the tensile test piece. In reality the material contracts in width and thickness. The stress calculated taking these contractions into account is called true stress. The true stress is calculated assuming that the volume of the tensile test piece does not change.

After the onset of necking, this is no longer true and therefore true stress can no longer be calculated easily. It can be determined by measuring the

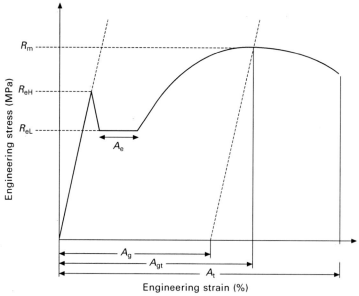

3.2 Schematic depiction of tensile curve with elongated yield point.

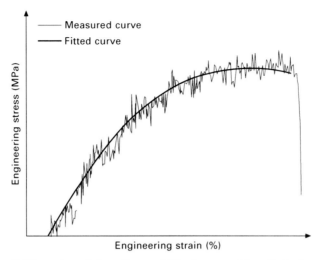

3.3 Experimental scatter and fitted curve of tensile test.

cross-sectional area at the location of the neck. True stress continues to increase with increasing strain as can be seen in Fig. 3.4 and represents the actual stress on the specimen up to the point of fracture. The engineering strain is defined as the ratio between the total deformation and the original dimensions. It is often expressed as a percentage.

$$e = (L - L_0)/L_0 \qquad\qquad [3.1]$$

In a test piece that is strained to twice its original length this would give a value of 100%. If the same test piece is now compressed to its original length then the engineering strain associated with this compression would be 50%. As this is not a meaningful measure in situations where a material is subjected to both compression and tension (as often in forming operations), we often use true strain, defined as the change in length compared with the instantaneous length and the total true strain as the summation of these.

$$\varepsilon = \Sigma((L_1 - L_0)/L_0) + (L_2 - L_1)/L_1 + \ldots) = {}_{L_0}\!\int^L \mathrm{d}L/L = \ln(L/L_0) \quad [3.2]$$

True strain provides the same numerical value in both tension and compression but of opposite sign. After the onset of necking the strain is no longer uniformly distributed over the total length and the equation cannot be used any more.

Plastic anisotropy: r-value

The anisotropy of a material has an influence on the distribution of strain. The strain ratio, or r-value, is the ratio of the true strain in the width direction to the true strain in the thickness direction (see equation 3.3).

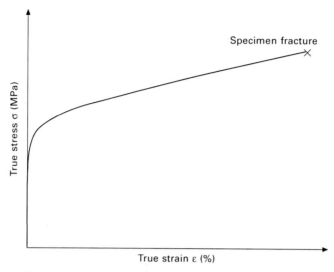

3.4 Schematic depiction of true stress-strain curve, with increasing stress until fracture.

$$r = \varepsilon_w/\varepsilon_t \qquad\qquad [3.3]$$

The strain ratio is related to the crystallographic texture of the material and varies in different directions in anisotropic materials. Therefore the strain ratio is determined in directions parallel (0°), transverse (90°) and at 45° to the rolling direction. The average plastic anisotropy, r_{ave}, can then be calculated (see equation 3.4).

$$r_{ave} = (r_0 + 2r_{45} + r_{90})/4 \qquad\qquad [3.4]$$

Yield loci

When uniaxial tensile testing is replaced by biaxial testing, the yielding of the material cannot be described by a yield point, but we need to measure the yield locus. The yield stresses for uniaxial tests and various biaxial tests can be plotted on a set of axes. These points can then be connected to give a graph called the yield locus. To fill in the missing points various criteria have been developed, such as Tresca and Von Mises. Each of these criteria will give a different shape of curve. Tresca assumes a straightforward linear relation between measured points. Von Mises assumes an elliptical shape, predicting higher yield stresses (see Fig. 3.5).

For more accurate analysis of forming behaviour (e.g., in finite element simulations) these models have been found to be inaccurate. Therefore more accurate criteria have been developed based on additional measurements, representing different stress-strain states (e.g., plane strain and pure shear),

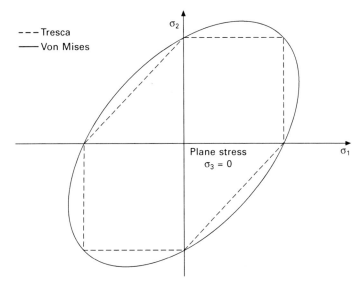

3.5 Yield locus according to Tresca and Von Mises.

to better estimate the shape of the curve. The Vegter-criterion, for instance, is based on the measurement of the uniaxial, plane strain, shear and equi-biaxial points. Beside the stress values in these points, the strain vectors are taken into account to construct the tangent to the yield locus. Between the reference points, a 2nd order Bezier interpolation is used where the hinge points are defined by the tangents of the two reference points in question (see Fig. 3.6).

The differences between the models are not trivial. Small changes may have a dramatic influence on finite element stamping predictions.

3.1.2   Work hardening

During cold forming work hardening will occur. Work hardening will cause the material to strengthen, thus increasing resistance against further deformation. Work hardening is caused by initiation of dislocations to accommodate plastic deformation. Increasing the amount of dislocations, without sufficient mechanisms to annihilate them, leads to decreasing mobility of these dislocations through the matrix. The dislocations become pinned by other lattice defects and start to repel each other, hindering further initiation of dislocations. Thus the material is hardened whilst residual formability is reduced.

Further deformation can only be achieved by increasing pressure, but this may cause damage in the material. Initially, damage will occur on a microscopic scale, which may lead to decreased performance of the parts (both

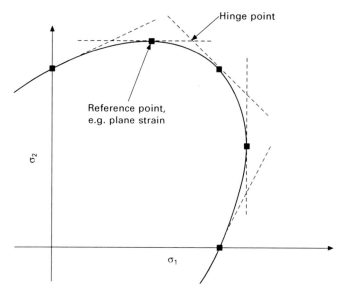

3.6 Yield locus according to Vegter.

affecting fatigue and crash). Excessive deformation will cause macroscopic cracks to occur.

Work hardening is an important feature during forming operations, because it strengthens the material. Although work hardening may ultimately limit (residual) formability, it is also beneficial. When a material is stretched local thinning increases strength through work hardening. The increased strength ensures that local straining becomes more difficult and that further straining occurs in surrounding regions. This increases the local formability of the material in critical areas.

Work hardening exponent: n-value

The work hardening exponent, or *n*-value, of a material is a measure for how quickly the material gains strength when it is being deformed. The *n*-value can be obtained from the slope of the true stress versus true strain curve in a tensile test, plotted on a logarithmic scale (see Fig. 3.7). The relationship stress and strain can be expressed in an equation:

$$\sigma = k\varepsilon^{n} \qquad\qquad [3.5]$$

where σ and ε are the true stress and strain and k and n are constants. Materials with high values of n show good formability. They can work harden sufficiently in critical areas to better distribute the strains over other areas thus reducing local build up of strains.

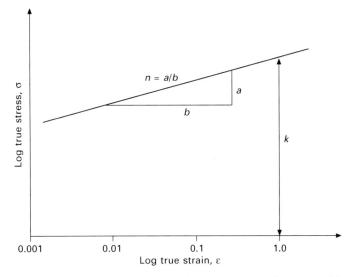

3.7 Logarithmic representation of true stress-strain curve with *n* the work hardening exponent.

Forming limit diagrams

Forming limit diagrams (FLDs) are graphical representations of the limits to forming; i.e., the major and minor stresses where local necking occurs. Although cracks are the ultimate limit in forming operations, local necking is usually considered undesirable.

FLDs can be generated by mapping the failure criteria on a graph of two axes representing major and minor strains. The major and minor strains can be measured using sheets with a grid. General techniques available for the grid involve regular patterns of circles, lines or dots, or randomly applied patterns. On straining the grid deforms. The major strain is defined as the strain in the direction of the maximum strain. The minor strain is the strain perpendicular to the major strain. The major strain is always positive and plotted on the vertical axis, while the minor strain is plotted along the horizontal axis. It can be positive or negative (see Fig. 3.8).

Most commonly, data is derived from forming tests with a hemispherical punch. Different stress conditions are simulated by changing the width of the (circular shaped) strips used for testing (see Fig. 3.9). The strips are drawn until failure occurs. The forming limit diagram can be determined for strain paths ranging from biaxial tension (as in stretch forming) to equal tension and compression (as in deep drawing).

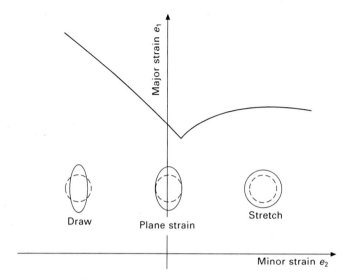

3.8 Schematic representation of an FLC (Forming limit diagram).

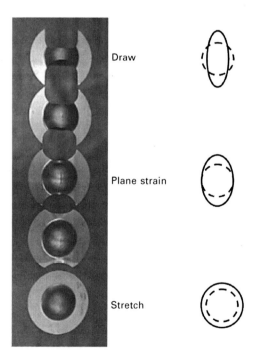

3.9 Experimental specimen for determination of forming limit diagrams.

3.1.3   Challenges

In addition to cracking, challenges that need to be considered when forming a work piece include:

- springback
- anisotropy
- earing and wrinkling
- thinning and necking
- strain rate behaviour.

Springback is the tendency of the material after cold working to (partially) return to its original shape, due to elastic recovery. Springback is primarily dependent upon the Young's modulus and yield stress of the material. Other factors also play a role, such as material thickness and the bend angle. Springback is especially important on the flanges of the work piece as these flanges are often critical to join the part to other parts of a construction. Severe springback may limit the options in joining operations. In spot welding an electrode force is used to push material together for welding. Although this compensates for some springback, severe springback may cause residual stresses after welding that can cause deformation or lead to decreased performance of the joints. In more advanced joining processes, such as laser welding and adhesive bonding, a good fit-up between parts is essential to allow joining. Severe springback may hinder the applicability of these processes or make the use of additional clamping necessary.

Anisotropy refers to differences in forming behaviour of a material in different directions. Especially when complex shapes need to be formed, material anisotropy may lead to undesirable consequences when the differences are not taken into account. Anisotropy is primarily caused by rolling operations needed to produce sheet materials. Hot rolled strip is reasonably isotropic, but cold rolled strip can show high levels of anisotropy. If a blank has been cold worked during previous operations, anisotropy can become even more pronounced. The amount of anisoptropy in the material is reflected in its r-value. An increase in isotropic behaviour leads to an increase of the r-value. An r-value of 1 represents isotropic material.

In 1948 Hill proposed an anisotropic yield criterion as a generalisation of the Von Mises criteria, assuming anisotropy in three orthogonal symmetry planes (so called Hill'48 criterion). In more complex materials, such as advanced high strength steels the issue becomes even more pressing as the orientation of grains is linked to variation in microstructure, such as the ferrite and martensite in dual phase steels. Although Hill'48 is still the most widely used model to describe anisotropy, the development of complex materials drives research into more accurate yield criteria, taking into account the anisotropic behaviour.

Earing and wrinkling influence the shape of the flanges of the formed part. Earing is caused by anisotropic behaviour and may lead to extra operations to remove excessive material. A measure for the amount of wrinkling can be calculated from the differences in planar isotropy:

$$\Delta r = (r_0 + r_{90} - 2r_{45})/2) \qquad [3.6]$$

A low value for Δr represents a low tendency for earing.

Wrinkling is a form of buckling that occurs due to compressive stresses in the flange and is generally classified as unacceptable. Wrinkling can be countered by increasing the blank holder force during stamping. There is a limit to this though, as too high a blank holder force will hinder material flow and cause fracture during forming.

Although thinning and necking are not as severe as fracture, they are generally undesirable as they may lead to decreased performance of the part. When complex shapes are to be formed some thinning of the material generally cannot be avoided, especially when the material has to be formed around sharp angles. Necking is generally associated with some internal damage of the material and is also the first step before fracture. For safe forming operations necking should be avoided.

Materials can be strain rate sensitive. Increasing strain rates change the mechanical response of materials. The yield strength of materials, R_p, increases with increasing strain rates as does the tensile strength, R_m. The plastic strain at fracture is reduced at increased strain rates. It is important to be aware of the influence of strain rates upon the forming characteristics of the material as the strain rate during tensile testing is usually much lower (e.g., $0.001\ s^{-1}$) than during forming operations (typically between $1\ s^{-1}$ and $10\ s^{-1}$). These considerations are mostly of importance for forming of steels as aluminium alloys are not affected that much by strain rate sensitivity at the strain rates usually experienced during forming operations.

3.2   Forming technology

3.2.1   Introduction

Depending on the material used (iron, magnesium, aluminium, carbon steel or stainless steel) several forming techniques are available to the automotive engineer to produce parts, such as casting, extrusion, forging and cutting. The overwhelming majority of parts in the body in white and chassis are produced by sheet metal forming. Both aluminium and steel sheet are used in forming. The choice between them is primarily dependent on the demands set for the part and the overall construction (e.g., strength, stiffness, performance in crash or surface appearance). Additionally there are demands on costs (both material and production), availability of manufacturing equipment and other engineering considerations (e.g., weldability and springback). Within this framework of boundaries an optimum combination between material and technique has to be found.

Sheet metal forming offers a wide range of techniques to the designer and engineer, such as:

- blanking
- stamping
- stretching
- bending
- roll forming
- hydroforming
- warm forming
- superplastic forming
- hot forming
- incremental forming
- explosive forming

which will be discussed in the following sections.

3.2.2   Forming techniques

Blanking

Before forming, coiled sheet material is de-coiled, straightened and cut into shapes suited for stamping. This process is called blanking. Blanking equipment can be integrated with stamping equipment into a continuous process line.

The geometry of blanks needs to be carefully tuned to the shape of the part. Obviously it is very undesirable to have blanks that are too small, but also blanks that are too large are undesirable. The need to remove excess material in later production stages requires additional measures in production to collect the waste material for recycling. Attention also needs to be given to the cutting operations, as rough edges on the blanks may lead to cracks during forming operations. Especially when more complex, higher strength materials are used, this may become an issue leading to high rejection rates. Edges may be ground after cutting to reduce the risks of occurrence of cracks, but this also requires an additional process step, leading to increased manufacturing costs.

Tailored blanks

When using a single blank; i.e., a blank made from one material, the design of the part has to meet the requirements of the most critical points. This complicates forming operations and may lead to parts that are overdesigned. To allow for a more precise application of material characteristics at critical points whilst saving material (and thus weight and costs), tailored blanks have been introduced for stamped parts.

Tailored blanks allow the engineer to make optimal use of material characteristics in forming. Instead of using a blank consisting of a single material that conforms to the maximum demands on the materials, tailored blanks allow for a careful use of optimal characteristics where they are needed. This can be achieved by welding sheets of different thickness together forming a single blank (thus allowing for reduction in weight), and/or by joining different grades (e.g., a high formable grade to a high strength grade) thus allowing for optimum performance of the formed part.

The introduction of a weld and its heat affected zone creates an element with different mechanical properties (e.g., softer in aluminium, harder in carbon steels) in the blanks, affecting local formability. Care has to be taken to ensure that the weld (which usually has limited forming characteristics) is not critically loaded during stamping operations. Accurate numerical analysis of the forming process of tailored welded blanks is used to design the tailored blanks so that the weld line is not subjected to excessive loading.

Tailor rolled blanks do not have the discontinuity of the weld in the blank. Instead they are manufactured by varying the thickness of the sheet material during cold rolling. The transition between thicknesses in tailor rolled blanks is gradual, which is a big advantage compared to tailor welded blanks. Also it is possible to produce a blank with several thickness transitions, without increasing the costs of manufacturing.

The use of tailored blanks complicates forming operations as the blanks have to be fed to the stamping press individually, whereas conventional forming presses can be incorporated in production lines where the uncoiling, straightening and stamping actions are aligned to allow continuous production. Additionally the welding of tailored blanks introduces an extra production step, which adds to costs of production. The advantages obtained in the application of the pressed part (savings in material and weight, as well as increased performance) often outweigh the extra costs.

Stamping

Stamping or press forming converts flat sheet material into three-dimensional shapes to produce complex parts and box sections of the body in white (BIW) of a car. Stamping is done by placing a blank of sheet metal between a male and a female die, respectively the punch and the die, in a stamping press. After insertion of the blank its outer edges are clamped. Clamping is dependent upon the amount of forming that is desired.

If deep sections are formed, additional material will be needed to flow from the outer edges into the die. This is referred to as (deep) drawing. In drawing operations the material is clamped using a blank holder. The blank holder serves to clamp the material with sufficient force to reduce the risk of wrinkles, but not so much that no material can flow into the die during pressing. After clamping the punch applies the stroke, which will cause the blank to deform.

The complexity of equipment and tools is completely dependent upon the geometry of the parts that need to be formed. Modern automotive designs require very complex parts with very deep sections (e.g., spare wheel wells) to be formed from very large blanks. These parts require the use of highly formable materials (mostly steels), and ever stronger equipment that can deliver higher press forces. Stamping presses can be driven hydraulically or mechanically. Hydraulically driven stamping presses apply a constant force during the stroke whilst in mechanically driven presses the force increases during the stroke. Usually complex parts are pressed in several stages.

Stretching

If no feeding of material is allowed during deformation the surface area of the sheet must increase during forming. Deformation is achieved by elongation and uniform decrease of the material thickness. Stretching requires the material to be clamped more rigorously than deep drawing. The *n*-value is generally considered a good measure of a materials formability in stretching.

Bending

In deep drawing the stresses and strains are the same all over the thickness of the material. A gradient will only occur where the material is drawn over a die radius. This is a secondary effect. In bending operations it is the aim of the operation. Die bending refers to a bend where the sheet is in contact with both the punch and the die to precisely define the shape of the work piece. V-bending, or free air bending, is used to set the material in an edge. The sheet is supported on two sides whilst a punch deforms the material in the middle. The final shape is influenced by the material properties as well as the die shape, as springback will occur. Bending usually requires simpler tooling than deep drawing and stretching.

Roll forming

During roll forming the sheet is bent by rolls in small steps into the desired shape. As such roll forming is bending with additional longitudinal strains, complicating operations (thus requiring more complex tooling). Roll forming operations are performed sequentially in a roll forming line, where sheet material is often fed directly from the coil into the equipment. The process is used to produce profiles. Tubes are also produced using roll forming. The tubes are closed by welding at the end of forming.

Hydroforming

Hydroforming can be used on sheets and tubes. Hydroforming of sheets can be either passive or active hydroforming (or hydro-mechanical forming). In passive hydroforming the punch forces the material against a body of fluid (usually water). As the pressure increases in the fluid the material is forced against the punch thus changing in shape. The absence of a die reduces friction leading to better surface quality and increased formability.

In active hydroforming the sheet is pushed against the punch by increased pressure prior to punch contact. The absence of friction during this bulging stage ensures straining in the middle of the part, increasing formability and improving appearance. Also the additional strain in the centre of the part increases hardening and therefore strength of the formed part. In tube hydroforming the tube is closed and filled with liquid. The liquid is pressurised and the pressure forces the tube to be formed. By controlling the pressure (mechanically or by flow control) a high shape accuracy can be achieved. Additionally the tube material itself can be pressurised by pushing the ends of the tube inwards, allowing for very high increases in formability.

Care has to be taken not to overload the weld and its heat affected zone, especially when high strength materials are used to produce the tube. Changes in strength and formability, due to the changes in microstructure in the weld and the heat affected zone may lead to failure when there is still sufficient residual formability left in the base material.

Hydroforming allows for reverse bending, which enables the material to flow in shapes that would be hard to form using traditional forming techniques.

Warm forming

The formability of aluminium sheet at room temperature is lower than for a typical deep drawing steel grade (about two thirds of mild steel). The Young's modulus is about one-third of steel, leading to more springback. The elongation is about half of that of steel. To improve formability the mechanical loading is combined with a thermal component, utilising the increased formability of aluminium at elevated temperatures. Parts of the tools can be heated or cooled to manipulate local flow behaviour.

Superplastic forming

Some aluminium (and titanium) alloys can be used for superplastic forming (i.e., elongation above 100%). The process was primarily used for aerospace applications, but aluminium closure panels have been produced by superplastic forming. To achieve superplastic properties the aluminium is alloyed with grain refiner elements (e.g., Cu and Zr). Thermomechanical treatment then assures a very fine grain structure (micrograins), needed for superplasticity. For superplastic forming the material is heated to around 500 °C, after which it is formed. Due to the elevated temperatures, the material is very weak which allows for the use of forming methods that are usually reserved for thermoplastic materials, such as blow forming and vacuum forming.

Strain rates need to be low so the cycle time in superplastic forming is long. Furthermore the addition of alloying elements increases material costs, though for automotive applications relatively inexpensive alloys with superplastic capabilities and good performance have been developed. Because of these limitations the application of the process has been limited to low-volume speciality vehicles.

Hot forming

For hot forming steel, sheet blanks are heated above the austenisation temperature (900 °C) using gas furnaces. After heating, the hot blank is placed in a stamping press. Because the material is austenitic during forming, high levels of deformation are possible. Due to the cooled tooling the material transforms to martensite during stamping. The result is a very strong part (> 1500 MPa) of high form accuracy (no springback).

The need to heat the blanks before pressing requires extra investment for production facilities. The cooling stage adds to the production time and the need to cool the tools requires more complicated equipment.

Incremental forming

In incremental forming the resultant shape is produced in very small increments. The circumference of the sheet is clamped and a pin (or a fluid jet) is used to form a small dent. The indenter is then moved over the surface of the sheet. The combination of the track of the indenter and the depth of indentation generate the profile of the work piece. Tools can be applied on a single side or on both sides of the sheet. A classical example of such a forming technique is panel beating.

Explosive forming

In explosive forming, a charge is used to apply a shock wave to the sheet material to force the sheet to deform against a die. The charge can be placed at a distance from the sheet, where the shockwave is carried to the sheet through a medium, or the charge can be placed directly on the sheet. The latter allows for higher pressures, but may also affect the sheet appearance.

3.2.3   Materials

Aluminium

The main reason to use aluminium sheet material is its combination of light weight and high strength. Though the strength of aluminium alloys is lower than that of steel, the light weight of the material allows for the use of thicker gauges that may compensate for the reduction in strength.

Aluminium alloys are classified according to their alloying components. Aluminium sheet materials are always wrought alloys, which are identified by a four digit number (see Table 3.1). The first digit in the alloy classification designates the group (i.e. the main alloy elements), the other digits designate the specific alloy. The alloy is usually followed by a code for the temper of the material, specifying the thermomechanical treatment during production (O for tempered materials, H for work hardened materials and T for heat treated materials).

Another important distinction between aluminium alloys is based upon their susceptibility to heat treatments. Heat treatable alloys contain elements like magnesium, silicon, copper and zinc, which are less soluble in aluminium at lower temperatures. As these elements are precipitated from the matrix, they form particles (e.g., Mg_2Si in 6xxx alloys) that strengthen the material. Precipitation can occur during heat treatment or at room temperature. The latter is called natural ageing. There is an optimum to precipitation hardening; if the heat treatment is continued for too long, the strength of the material decreases and the material is said to have overaged. Some alloys, specifically

Table 3.1 Aluminium alloy classification

Group	Main alloy elements	Heat treatable
1xxx	Al	No
2xxx	Al-Cu	Yes
3xxx	Al-Mn	No
4xxx	Al-Si	No
5xxx	Al-Mg	No
6xxx	Al-Mg-Si	Yes
7xxx	Al-Zn-Mg	Yes
8xxx	Al-other	

6xxx (AlMgSi) and 7xxx (AlZnMg) continue to age harden for long periods, which limits their shelf life in terms of formability.

The 1xxx group consists of commercially pure aluminium alloys. They are highly weather and corrosion resistant and show good formability (and weldability), but their low strength makes them unsuitable for body and chassis applications. They can be used for other parts such as light reflectors.

The 3xxx group alloys are stronger whilst retaining high formability (and weldability) at low costs. Their strength levels are generally not sufficient for application in structural components, but a lot of 3xxx sheet material is applied indirectly in the form of aluminium brazing sheet (3xxx cladded with 4xxx), the main material to manufacture heat exchangers for automotive applications.

Alloys in the 5xxx group are generally stronger than alloys in the 3xxx group as magnesium is a more effective strengthener than manganese (which is also added to the 5xxx alloys to further increase work hardenability). After work hardening 5xxx alloys are strong and suited for structural applications.

The 6xxx alloys are heat treatable after forming, increasing their strength through precipitation of Mg_2Si. High form accuracy and surface appearance make these alloys suitable for production of profiles. Special precautions may be needed when these alloys are welded because of decrease in strength through the addition of heat.

The 7xxx alloys can be precipitation hardened to achieve very high strength levels; in fact the highest strength levels achieved in aluminium. Natural ageing causes these alloys to increase in strength and decrease in ductility over time.

Steel

The main driver for the development of steel grades has been the demand for stronger steels. Strength alone does not make for a steel grade that is suitable for automotive applications. The steels need to posses good formability and weldability, and depending on the application of the part, excellent corrosion resistance, crash worthiness, dent resistance and performance in fatigue. The relation between strength and formability for steels is often depicted in the well known banana-curve (see Fig. 3.10) where tensile strength is related to formability (usually elongation)

Low alloyed steels, consisting of very clean material often exhibit excellent formability. Examples of these steels are interstitial free steels and forming steels such as DC-grades. These steels are limited in strength though, due to the low levels of alloying (specifically carbon).

Steels used for exposed panels need to be highly formable with very good surface appearance. To satisfy this requirement their yield strength needs to be kept low. The exposed panels not only need to be structurally

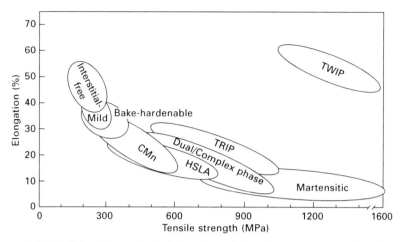

3.10 The 'steel banana' depicting the relation between strength and formability.

rigid (adding to the stiffness of the construction) but also have good dent resistance to withstand impact of small stones, branches, fingers, etc. To satisfy this requirement the material needs to be strong. Bake hardening steels were developed to satisfy both requirements. Bake hardening steels are alloyed with carbon and nitrogen in solid solution, keeping the yield strength low prior to press forming. After press forming the parts are painted and the paint is hardened at elevated temperatures (typically 170 °C for 20 minutes). During this paint baking cycle the carbon and nitrogen diffuse to the dislocations generated during press forming. The precipitates pin the dislocations thus increasing the strength of the material.

Higher strength can be achieved in steels by increasing the carbon content, but this affects both formability and weldability. Rephosphorised steels were developed to increase strength levels with good formability. Strength levels are increased in Rephos steels by solution hardening with phosphorus and manganese. These steels perform well in stretch and deep drawing modes, but there are some issues with weldability.

HSLA (high strength low alloyed) steels were initially developed to allow for high formability, without affecting weldability. Increased strength levels are achieved by micro-alloying low carbon steel sheet. The fine grained ferrite-pearlite microstructure is strengthened by titanium, vanadium, niobium or chromium carbonitrides. The ferrite in HSLA steels allows for considerable formability, and weldability is maintained because of low alloying levels.

The demand for stronger steels led to the development of advanced high strength steels such as DP (dual phase) and TRIP (transformation induced plasticity) steels. Dual phase steels consist of a microstructure of ferrite and martensite, allowing good formability (due to the ferrite) with increased strength levels (due to the martensite). TRIP steel consists of a microstructure of ferrite, bainite and residual austenite. The inclusion of an austenitic phase allows for additional formability by transformation to martensite. Thus the forming operation adds to the strength of the formed part.

Advanced high strength steels show better elongation than can be achieved by traditional steels of similar strength levels, that make use of solid solution, precipitation and transformation hardening. Advanced high strength steels are especially suitable for parts that need to have high crash worthiness, as their high formability makes them suitable for producing complicated shapes, whilst the ferritic component is beneficial for their energy absorbance at high strain rates.

The need for increased alloying (e.g., with carbon and phosphorus) can cause issues with the weldability of these steels. Further development of more complex advanced high strength steels, such as HyPerform (combining elements of DP and TRIP steel) and quenched and partitioned steels, allows for increased formability combined with higher strength levels, whilst retaining good weldability.

Hot stamping or boron steels have increased hardenability. The increased carbon content of these steels (compared to traditional forming grades), combined with the addition of boron ensures the transformation to martensite after stamping. After heating, the fully austenitic materials exhibit increased formability, allowing deep drawing levels that cannot be achieved with ferritic steels of similar alloying levels. After stamping the fully martensitic materials ensure very high strength levels and form accuracy.

Although hot stamping steels allow a break from the traditional relation between formability and strength, the requirements for forming operations and the investments needed to use these materials to their full effect are a drawback. Additionally, there are issues with appearance, corrosion and the fact that these materials possess little residual formability after stamping, making them unsuitable for critical applications such as crash resistant parts. Therefore completely new approaches in steel manufacturing are under development to produce materials that do not behave like traditional carbonsteels. TWIP (twinning induced plasticity) steels are fully austenitic steels that combine high strength with good formability (due to the formation of twins). The increased levels of alloying (e.g., manganese) increase the price and may complicate other aspects of automotive manufacturing (welding, recyclability, *et cetera*).

Stainless steel

Stainless steels are steel alloys with a minimum chromium content of 10.5 wt% and a maximum carbon content of 1.2 wt%. This ensures the presence of a self-healing oxide layer on the surface of the material, providing the corrosion resistance. The chemical composition determines the metallurgical properties of stainless steels. There are four main families of stainless steels:

1. austenitic stainless steels (Fe-Cr-Ni, with C < 0.1 wt%)
2. ferritic stainless steels (Fe-Cr, with C < 0.1 wt%)
3. duplex stainless steels (Fe-Cr-Ni)
4. martensitic stainless steels (Fe-Cr, with C > 0.1 wt%).

The different families have different mechanical properties. The martensitic stainless steels have high strength, but low ductility, whereas the austenitic stainless steels have lower strength but much higher ductility. The ductility of ferritic and duplex stainless steels lie between those of martensitic and austenitic alloys, with ferritic stainless steels having similar strength levels to the austenitic alloys. The strength levels of the duplex steels are higher than those of the austenitic alloys, but lower than those of the martensitic materials.

Two mechanisms of work hardening operate in austenitic and duplex stainless steels. Apart from work hardening due to the movement of dislocations, as described before, the austenitic phases may transform during cold working into a hard martensitic microstructure of limited ductility. The combination of high formability with the high strength of the resultant product, offers possibilities for weight saving and increased crash performance.

3.2.4   Tribology

Apart from the bulk of the material, the sheet metal surface also plays a role in defining material performance in sheet metal forming. Obviously, surface appearance after forming is important for many parts, but the quality of the surface may also be critical for a successful forming process. The interaction of the surface of a material with forming tools is a subject of the field of tribology. Tribology during forming operations is concerned with two aspects: friction and wear.

Two types of wear can be defined, adhesive wear and abrasive wear. Abrasive wear causes loss of tool material, thereby reducing tool lifetime and increasing manufacturing costs. Adhesive wear is related to the transfer of sheet material to the tool surface and this can lead to galling and surface damage on formed parts and to tool pollution, thereby increasing scrap rates and tool maintenance costs. The increasing use of higher strength sheet materials in the automotive industry means that both kinds of wear are becoming more of an issue. Tool material selection is becoming more important and the use of more advanced, but also more expensive, tool steel grades is becoming more attractive due to extended tool lifetimes.

Friction controls material flow into the die during forming operations. High friction should be avoided as this can restrict material formability and cause fracture in a part however, low friction can also cause problems, because a certain control over material flow is lost.

Frictional performance and wear behaviour are not material properties but system phenomena dependent on material parameters (tool material, sheet surface coatings/quality, lubrication, etc.), process parameters (including forces) and part geometry. Careful selection of these parameters can be used to optimise friction and wear behaviour and ensure a successful process.

Sheet surface quality is affected by the surface texture. The surface of sheet material should have some roughness, as this is required for lubricant retention, paint adherence and appearance. The surface roughness is determined by the rolling operations during the manufacture of the sheet material. Hot rolled materials generally have a higher surface roughness than cold rolled materials. The quality is dependent upon the surface quality of the rolls during the manufacturing operations, which deteriorates over time. Therefore rolls are regularly replaced. Surface texture is dependent upon the texture of the surface of the rolls. They are roughened using a variety of techniques, such as blasting (SBT or shot blast textured), sparking (EDT or electro discharge textured), electron beaming (EBT) or grinding (MF or mill finish).

Sheet material is often coated to improve corrosion resistance. For steel, zinc coatings are most commonly used. Depending upon the process used to apply the zinc, sheet steel can be galvanised (GI), galvannealed (GA) or electro-zinc coated (EZ). Each of these coating types has unique tribological properties. Additional layers of coating can be added depending on customer demands, e.g., to provide passivation. Coated sheets are oiled to protect the material during coiling and further handling and sometimes to provide lubrication during forming.

3.3   Modelling

Finite element analysis (FEA) has become one of the major tools of mechanical engineers in design, engineering and testing, saving enormous amounts of time, money and material. Since the development of the finite element method (FEM) in the aerospace industry by Boeing and Bell in the United States and Rolls Royce in the United Kingdom in the 1950s, FEM has found its way into automotive engineering. There it has been applied with great success for structural analysis, design, and engineering, such as forming operations.

The basic idea of FEM is to divide a body into finite elements, often just called elements, connected by nodes, which are subjected to a set of (virtual) processing and material conditions to obtain an approximate solution. The procedure of computational modelling using the FEM broadly consists of four steps:

1. modelling of the geometry
2. meshing (discretisation)
3. specification of material properties
4. specification of boundary, initial and loading conditions.

These steps are carried out during pre-processing, after which a solver is used to calculate a solution. The solution can then be analysed using post-processing tools available within the software.

Simulation of forming operations is generally aimed at manufacturability; i.e., how to make a part. FEA can be used to assess critical areas during forming, where excessive stress or strain levels may lead to necking or failure. If problems are to be expected, FEA can give an insight into the optimal solution: changing the tooling, changing blank holder forces, application of lubricants or application of different grades of material. It is used to predict (and counteract) wrinkling and springback, and to assist the introduction of new grades of materials in an existing design.

If these issues had to be tackled using experimental data solely, the costs of part design and engineering would increase tremendously. Even trying to approach a more ideal solution would require an enormous amount of testing, with associated costs of manpower and equipment as well as material. Introduction of new materials poses an extra challenge due to the fact that when new promising materials are developed their availability may be limited and their price still high. If it was not for FEA, many of the great advances that have been made in the design of vehicles in terms of increasing safety whilst limiting weight, would have been very difficult to achieve. It is precisely because the costs of tooling and equipment in forming operations can be so high that the investigation of an ideal solution via simulation is profitable in forming technology (although modelling and simulation requires a substantial investment in people and software).

The first FEM software platforms developed were NASTRAN at NASA and ANSYS at Westinghouse Electric. Since then a wide variety of software packages has become commercially available, such as Abaqus, ANSYS and LS DYNA as well as software developed for academic purposes, such as DiekA. These software packages can be used to develop specific models used for simulations, but solutions have been developed and commercialised specifically for forming applications, such as PAMStamp and AutoForm. All of these packages have unique advantages and disadvantages and the ideal solution for a problem depends on the goal that needs to be achieved, in combination with the resources available (hardware, people, available data).

Though FEM is well established, it is still very much under development. Aided by the availability of ever more powerful hardware and software, more and more complex models are set up. For simulations to be worthwhile for application, the effort put into the modelling needs to be less than an equivalent experimental approach would require. Therefore a lot of effort is spent on reducing calculation times. Another field of development is on the modelling tools. Improvements in meshing techniques, elements capable of carrying specific information (e.g., for welds) and comparison and visualisation of results are continuing.

The introduction of new forming techniques has led to the development of equivalent models. In recent years tailored welded blanks and hydroforming models have been developed and introduced with great success. Currently a lot of effort is aimed at the development of hot forming models, incorporating mechanical forming models with thermal phase transformation models.

Through process modelling, linking forming operations with other aspects of automotive manufacturing are in full development. On the one side this encompasses the introduction of welds (such as found in tailor welded blanks and tubes) and a drive to link the thermomechanical history of sheet materials (as experienced by the material during production) with forming operations (especially of interest for the different tempers in aluminium sheet). On the other side there is a drive to link forming operations with subsequent steps, such as welding (assembly) and crash.

The challenges lie in the transfer of mesh and data from one software platform on to another mesh and (often) software platform suitable for follow up simulations. Often the size of elements and nature of elements differs. For instance to transfer a mesh from forming simulations to a mesh suitable for welding simulations requires not only a difference in size and density, but also a transformation from shell elements to solid elements. Additionally there is a difference in the data that is required to perform meaningful simulations; forming simulation requires only mechanical data, while welding simulations require input data on phases and thermal characteristics (e.g., heat capacity and conductivity).

Some successes have been achieved, especially in forming to crash simulation, but there is still a lot of work to be done. Without doubt, new challenges will arise as soon as these are addressed.

3.3.1   Material data for finite element calculations

FEA has proved its worth in mechanical engineering, and few would doubt its usefulness today. Its popularity in industry can be gleaned by the fact that over $1 billion is spent annually in the United States on FEM software and computer time (not just for automotive applications). However, the value of its predictions is very much dependent on the operator of the software. Though the software can generally be trusted to return the right results, the way a model is set up and the analysis of the results by the engineer determines the validity of the results of simulation. In forming simulation a major part of the input is mechanical material data.

In principle, measured material data could be used directly to engineer forming operations, but they are more often used to facilitate numerical simulation of forming. Measured data usually has to be fitted to be used in simulations, essentially building a model of the material behaviour. Measured data contains scatter. Scatter can lead to numerical irregularities. The stress strain curve is therefore fitted to a (second order) polynomial. This curve still needs some adaptations to be suited to most commercially available finite element software.

Usually, measurements are made in three directions: in the rolling direction (0°), transverse direction (90°), and at 45°. This will often result in three different curves. This is solved by using the curve in the rolling direction and having the curves in the other directions following from the combination of the hardening curve with the yield locus. There is a long list of strain hardening laws available in FEM codes (e.g., Ludwik–Nadai), but they all give an analytical equation that describes the relation between stress and strain. More complex equations incorporate strain rate hardening and temperature (e.g., Bergström–van Liempt). Temperature is increased by plastic work during deformation and has an influence on strain hardening.

The actual fitting of strain hardening curves to the experimental data is done in four steps:

1. convert experimental data to true stress – true plastic strain
2. prefitting
3. fitting
4. generating curves from parameters.

Prefitting prepares the experimental data for better fitting by noise reduction, interpolation and temperature calculation. In the fitting step the parameters for the strain hardening laws that will be used are determined. It is essential to ensure that coherent dimensions are used when preparing data for FEM software.

How a material is represented in a model is primarily dependent upon the software platform. Depending on the FE package, materials can be implemented using the parameters obtained or curves generated from these parameters. Not all material data are always known and measurement is not always possible. In that case educated guesses and interpolation of available data points may sometimes be necessary. Here again the influence of the engineer operating the software is of critical importance. Materials have long ceased to be purely physical entities and their virtual representation deserves the utmost care to enable successful forming simulations.

3.4   Economic considerations

In recent years there has been a drive in automotive design and manufacturing towards increased safety and fuel efficiency.

Increased safety is driven by legislative pressures and customer demand. Increasingly, regulators have pressed for increased performance in occupant and pedestrian safety, which have been addressed by improved design and increasing the strength and crash performance of materials used. Meeting regulators demands is often not sufficient to be successful in the market, as customers will prefer vehicles with top crash ratings. Not meeting the highest rating, may cause serious damage to sales figures, as the introduction of an affordable SUV on the European market in 2005 illustrated, when bad publicity surrounding initial crash results caused sales figures to collapse overnight.

Increased fuel efficiency is driven by regulations and economics. Regulators press for improved fuel efficiency with the aim of reducing the emission of greenhouse gases. Although this is also an important point for the general public, the temptation of big and fast cars has usually been too strong to resist. Increasing fuel prices have however proven to be efficient drivers for an increased demand for fuel efficiency. Fuel efficiency can be addressed by improving the overall performance of the vehicle, especially the power train, but there are significant gains to be made in the body and chassis. Most prominent of these is weight reduction.

Weight savings in the body and chassis can be made by improved design and production. Optimal design uses materials efficiently, where their impact is most beneficial for the overall performance of the vehicle. Application of alternative welding techniques may allow for a reduction of flange width on the parts, which will lead to savings in weight. Precise cutting of well designed blanks can do the same. Two further approaches have been proven to have the biggest impact on weight savings: substitution of materials and downgauging.

Substitution of steel by aluminium (and possibly magnesium) has obvious advantages for weight reduction. Aluminium is especially attractive for large parts that need not resist heavy loads, such as closures and roof panels, and offers an additional benefit of lowering the centre of gravity of the vehicle, contributing to driving comfort. The impact of a substitution of steel by aluminium on forming operations is evident, e.g., new equipment is required (especially tooling). The formability of aluminium is generally lower than that of steel, so for more complex parts warm forming may be required (especially if surface appearance is important). Issues also arise when aluminium parts need to be joined to steel parts, and the difference in heat expansion may cause problems during the paint baking cycle. Additionally, recycling may be an issue. Not only when the vehicle has reached its end-of-life, but also in a factory environment where for instance scrap metal routes for steel and aluminium will have to be separated.

Downgauging of sheet material is a very effective approach to reduce weight. Again there are consequences for forming operations. Process adaptations are needed, e.g., on tooling and blankholder force. As stress is a property determined by thickness, designs may need to be changed to avoid necking and failure. There will be critical locations where downgauging is not an option, because the performance of the part requires a certain material thickness, e.g., to attain a certain stiffness. The solution may be found in increasing the number of parts with varying thickness to provide strength and stiffness only where it is needed. Additional process steps will be required as more parts will have to be assembled, increasing handling and assembly operations. A solution could also involve the use of tailor welded blanks that are assembled prior to forming. Downgauging can also have a drawback; when thinner materials are used, the stiffness of a construction may decrease.

The stiffness of the construction is an important parameter determining driving comfort. Stiff constructions add considerably to driving comfort. They also contribute to fuel efficiency as losses in mechanical energy are reduced. Though improved design (e.g., incorporating continuous joints) can make up for a loss of thickness, there is a limit. Driving comfort is of course a considerable selling point.

The combination of safety and weight savings has led to the introduction of new materials in automotive manufacturing, where an increase in strength and energy absorption of materials has allowed manufacturers to downgauge without the loss of performance. Bake hardening steels have allowed manufacturers to reduce the weight of closures without loss of dent resistance and appearance. Advanced high strength steels have been introduced to reduce the weight and improve the performance of crash parts in the construction. Hot stamping steels have been introduced to reduce weight with considerable improvement of performance for side impact parts. The introduction of these materials requires investment in forming operations, most notably in the case of hot stamping steels (requiring considerable investment in equipment).

The most important parameter when considering what forming techniques and materials are to be used when producing a vehicle is the manufacturability; the actual making of a car. Within the window of possibilities that originates from this, choices have to be made about the optimum mode of production. Obviously, costs are always a very important issue, but choosing simply the cheapest options will not necessarily lead to the ultimate goal: maximum profit.

Materials with improved properties are generally more expensive than traditional low carbon steels, but the savings can be considerable. As metals are priced per tonne, downgauging can lead to cost reduction as less metal is needed to produce parts. Additional savings can be made because less scrap is produced.

The need to change forming operations requires investment, whether that is cost beneficial is dependent on a number of parameters, the most important often being production volume. Steel is cheaper than aluminium, but the benefits gained from a multi metal approach in the performance of a vehicle may outweigh the initial monetary savings as cars are not just machines, they have an emotive quality that has a cost benefit to the manufacturer. Customers have consistently been found to be (sensibly) reluctant to sacrifice comfort, safety and performance for price alone.

3.5   Bibliography

3.5.1   Formability

- D. Banabic, F. Barlat, O. Cazacu & T. Kuwabara: Advances in Anisotropy and Formability; *Int. J. Mater. Form.* 3; pp. 165–189; 2010.
- J.L. Bassani, J.W. Hutchinson & K.W. Neale: On the Prediction of Necking in Anisotropic Sheets; Reprint from *Metal Forming Plasticity*; IUTAM Symposium Tutzing/Germany; August 28–September 3, 1978; Ed. H. Lippmann; Springer-Verlag; 1979.
- W.D. Callister: *Materials Science and Engineering*; 7th ed; John Wiley & Sons; ISBN-13: 978-0-471-73696-7; 2007.
- ISO 6892: Metallic Materials – Tensile testing – Part 1: Method of test at ambient temperature.
- ISO 10275: Metallic Materials – Sheet and strip – Determination of tensile strain hardening exponent.
- ISO 10113: Metallic Materials – Sheet and strip – Determination of plastic strain ratio.
- R. Hill: A theory of the yielding and plastic flow of anisotropic metals. Proceedings of the Royal Society of London; Series A. Mathematical and Physical Sciences. Volume 193, No. 1033, pp. 281–297, May 27, 1948.

3.5.2   Forming Technology

- M.H.A. Bonte: *Optimisation Strategies for Metal Forming Processes.* PhD-thesis; University of Twente, The Netherlands; ISBN: 978-90-365-2523-7; 2007.
- A.H. van den Boogaard: *Thermally Enhanced Forming of Aluminium Sheet – Modelling and Experiments.* PhD-thesis; University of Twente, The Netherlands; ISBN: 90-365-1815-6; 2002.
- W.C. Emmens: *Tribology of Flat Contacts and its Application in Deep Drawing.* PhD-thesis; University of Twente, The Netherlands; ISBN: 90-3651028-7; 1997.
- A.M.H. Hadoush: *Efficient Simulation and Process Mechanics of Incremental Sheet Forming*; PhD-thesis; University of Twente, The Netherlands; ISBN:978-90-365-3052-1; 2010.

- S. Kurukuri: *Simulation of Thermally Assisted Forming of Aluminium Sheet*; PhD-thesis; University Twente, The Netherlands; 2010.
- R. Pearce: *Sheet Metal Forming*; IOP Publishing; ISBN 0-7503-0101-5 1991.

3.5.3   Materials

- H. Bhadeshia & R. Honeycombe: *Steels – Microstructure and Properties*; 3rd edn; Butterworth-Heinemann; ISBN-13: 978-0-750-68084-4; 2006.
- D. Bhattacharya, H. Guyon, A. Belhadz & N. Fonstein: New Developments in Advanced High Strength Steels for Automotive Applications; World Automotive Materials Meeting; 2005.
- B. van Hecke: *The Forming Potential of Stainless Steel*; 1st edn; Materials and Applications Series, Volume 8; Euro Inox 2006; ISBN 978-2-87997-211-4; 2006.
- H. Hofmann, D. Mattissen & T.W. Schaumann: Advanced cold Rolled Steels for Automotive Applications; *Steel Research International* 80 (1); 2009.
- H. Karbasian & A. E, Tekkaya: A Review on Hot Stamping; *Journal of Materials Processing Technology* 210; p.p. 2103 – 2118; 2010.
- J. Kell: Automotive Advances; *Materials World*; April 2008.
- D.E. Kim: High Strength and High Elongation Dual Phase Galvannealed Steel Sheets Development; *SEAISI Quarterly Journal* 37(2), 2008.
- D.T. Llewellyn: *Steels: metallurgy and applications*; Butterworth-Heinemann Ltd; ISBN 0-7506-2086-2; 1992.
- P.D. Marchal: *Aluminium in dunne plaat en buis*; Tech-Info-Blad nr. TI.04.21; Vereniging FME-CWM; June 2004 (in Dutch).
- P.D. Marchal: *Hoge Sterkte Staal in dunne plaat en buis*; Tech-Info-Blad nr. TI.04.18; Vereniging FME-CWM; January 2004 (in Dutch).

3.5.4   Modelling

- R.A.B. Engelen: *Plasticity-induced Damage in Metals – Nonlocal modelling at finite strains*. PhD-thesis; Technical University Eindhoven, The Netherlands; ISBN: 90-771-7216-5; 2005.
- J. Fish & T. Belytschko: *A First Course In Finite Elements*; John Wiley & Sons Ltd; ISBN 978-0-470-03580-1; 2007.
- G.R. Liu & S.S. Quek: *The Finite Element Method a Practical Course*; Butterworth-Heinemann; ISBN 0-7506-5866-5; 2003.
- S.S. Rao: *The Finite Element Method in Engineering*; Elsevier Science & Technology Books ; ISBN: 0-7506-7828-3; 2004

4
Nanostructured steel for automotive body structures

Y. OKITSU, Honda R&D Co. Ltd., Japan and
N. TSUJI, Kyoto University, Japan

Abstract: In this chapter, the results of studies on the application of nanostructured steel sheets to automotive body structures are introduced. A new route to fabricating nanostructured/ultrafine ferritic microstructures without severe plastic deformation is presented. The preparation and evaluation of two kinds of steel sheets, ultrafine-grained (UFG) ferrite-cementite (FC) steel and UFG multi-phase (MP) steel are also presented and discussed. The UFG-FC steel showed large strain rate sensitivity in flow stress, while the UFG-MP steel showed a good combination of high strength and large work hardening rate, which improved the dynamic collapse properties of hat columns. It is shown that UFG steels would help further weight reduction of automotive body structures.

Key words: ultrafine grain, steel sheets, multi-phase, strain rate sensitivity, dynamic collapse.

Note: This chapter was first published as Chapter 22 'Applying nanostructured steel sheets to automotive body structures' by Y. Okitsu and N. Tsuji in *Nanostructured metals and alloys*, ed. Sung H. Whang, Woodhead Publishing Limited, 2011, ISBN: 978 1 84569 670 2. It is reproduced without revision.

4.1   Introduction

In recent times, reducing the body weight of automobiles has been considered one of the important criteria in auto manufacturing for reducing CO_2 emissions as well as improving the fuel efficiency of automobiles. Steels have been the most popular and indispensable material for automotive body structures in the past, and recently in particular, various kinds of high-strength steels (HSSs)[1] have been developed and applied to the body structures, which has contributed to a weight reduction. In order to achieve further weight reduction, new HSSs are required. As described in the next section, nanostructured or ultrafine-grained (UFG) steels are expected to become one of the new HSSs.[2] However, it is still difficult to apply UFG steel sheets to automobile body parts. This is owing to the limited dimensions of UFG steels fabricated by severe plastic deformation (SPD) processes such as equal-channel angular extrusion (ECAE),[3] high-pressure torsion (HPT),[4] and accumulative roll bonding (ARB),[5–7] which require very high plastic strain.

57

Recently the authors showed a new route to fabricate UFG steel sheets through conventional rolling and annealing procedures,[8] which made it possible to overcome the dimensional problem, and evaluate the properties and performances required for automotive body structures systematically. Based on this processing route, UFG steel sheets with various microstructures such as ferrite-cementite (FC)[8,9] and multi-phase (MP) structures[10] were fabricated.

In this chapter, future trends of automotive body engineering and demands for UFG steels are described first. Secondly, the new route to fabricate UFG steel sheets without SPD and the properties of fabricated UFG-FC steel sheets are shown. Next, UFG-MP steel sheets developed in order to improve the work hardening rate of UFG microstructures are described. Finally, there is an illustration of the crash-worthiness of UFG steel sheets, which is one of their important properties as body materials.

4.2   Potential demand for nanostructured steels for automotive body structures

Nowadays, improving the fuel efficiency of automobiles is a very important subject, which comes from global demands for reducing CO_2 emissions. For this purpose, it is necessary not only to improve the fuel efficiency of engines, but also to reduce body weight, rolling drag force, friction, etc. However, the actual body weight of automobiles has increased mainly due to the demands for improving the crash-worthiness of body structures during a collision. For this reason, considerable efforts have been made to suppress any weight increase. The applications of HSSs in automobiles have increased significantly since the 1990s.[11] As a result, more than 50% of body parts are made of HSSs[12] in recent cars. Although low-density metals such as aluminium and magnesium are also applied, low-carbon steels are still widely used for manufacturing various kinds of cars due to their high cost performance. The ULSAB (UltraLight Steel Auto Body) project showed that a significant weight reduction could be achieved by applying AHSSs (advanced high strength steels) and advanced manufacturing methods. A review of such projects is presented in the appendix to this chapter.

In order to achieve further weight reductions by extending the application of HSSs, new steels with new microstructures and superior properties have been widely studied.[13] An example is high-Mn TWIP (twinning-induced plasticity) steels,[14,15] which have austenite microstructures and a combination of superior tensile strength and elongation. This steel is classified as 'second-generation AHSS'.[2] A recent target, so-called 'third-generation AHSSs',[2] is steels with intermediate elongation between conventional HSS and TWIP steels but fewer alloy elements than TWIP steels. Grain refinement down to submicrometer grain sizes is thought to be one of the important microstructural controlling methods that will contribute to third-generation AHSS.[2]

Up to now, the improvement of automotive steels has been discussed in terms of formability. However, in order to achieve further weight reduction, other points of view are being considered. Generally the parts in an automotive body structure are classified roughly into two groups having different functions. One is the parts in crushable zones, i.e. the front and rear frame sections, which deform heavily in car collision and should absorb impact energy. The other is the parts making up the passengers' cabin, which deform little and prevent the passengers from injury during a collision. Ultra high strength steels (UHSSs) with tensile strength of over 1000 MPa are applied to the parts for making up the passengers' cabin.[16] However, if UHSSs are applied to the parts in the crushable zones aiming for further weight reduction, other mechanical properties must be preferentially considered. For example, when UHSSs are applied to the crushable parts, their deformation mode tends to become unstable, which reduces the efficiency of energy absorption.[10] This is caused by the decreased work hardening rate of the material already highly strengthened.[17] The authors[10] have shown using lab-fabricated UFG-MP steel that a large work hardening rate (n-value) was effective in achieving a stable deformation in axial collapse. In the following sections, the practical results of evaluating microstructures, mechanical properties and crash-worthiness using fabricated UFG steel sheets are shown and discussed.

4.3   Fabricating nanostructured low-C steel sheets

4.3.1   A new route to fabricate nanostructured steel sheets without severe plastic deformation

It has been already reported that UFG ferritic steels have superior properties such as high strength,[18] high fracture toughness at around −190 °C[19] and high strain rate sensitivity in flow stress.[20,21] Severe plastic deformation processes are generally applied in order to fabricate nanostructured or UFG steels. However, they do not seem to adapt to conventional mass production routes for steels. In this section, firstly, the concept of the new route for fabricating UFG steel sheets without SPD is described. The distinguishing feature of the process is conventional rolling of a duplex microstructure composed of soft ferrite and hard martensite. The key factor in the process is strain distribution between soft and hard phases. When such a duplex microstructure is deformed, a larger strain is introduced in the soft phase (ferrite), while a strain that is not so heavy but sufficient for final microstructural refinement is introduced in the hard phase (martensite). As a result, a number of recrystallization nuclei form uniformly from both ferrite and martensite, and uniform UFG microstructures are obtained after subsequent annealing.

Table 4.1 shows the chemical composition of the UFG-FC steel studied. This steel is made by adding Nb and B to a low C steel. Figure 4.1 shows a schematic illustration describing the fabricating process and conditions. The fabricating process consists of conventional hot-rolling, cold-rolling and annealing as indicated in Fig. 4.1.

An ingot was hot-rolled to a thickness of 6.8 mm at austenite region, air-cooled to 540 °C, which corresponded to the intercritical region of ferrite and austenite, and water-cooled to room temperature for obtaining a duplex microstructure of ferrite and martensite. The hot-rolled sheets were cold-rolled at room temperature with lubricant. Specimens having various total cold-rolling reductions from 85% to 94% were prepared. The cold-rolled sheets were annealed at various temperatures ranging from 525 °C to 700 °C for 120 seconds in a salt bath. Annealed sheets that were 150 mm wide and 0.4 to 1.0 mm thick were obtained. Microstructural observations were carried out by optical microscopy and scanning electron microscopy (SEM).

An optical micrograph (OM) of the hot-rolled steel sheet observed from the transverse direction (TD) of the sheet is shown in Fig. 4.2. The hot-rolled sheet showed a duplex microstructure composed of a ferrite matrix (bright region) and martensite islands (dark region). The mean intersect lengths along the normal direction (ND) of both regions, measured using the OM, were 5.55 µm in the ferrite and 2.46 µm in the martensite. The area of martensite was 42% of the whole. Transmission electron microscopy (TEM) observation has confirmed that the martensite present was 'lath-martensite' including a high density of dislocations. Cementite particles were not observed in the as-quenched martensite.

Table 4.1 Chemical composition (mass%) of UFG-FC steel

C	Si	Mn	P	S	Al	Nb	B	N
0.10	0.01	2.00	0.002	0.0013	0.035	0.022	0.0015	0.0007

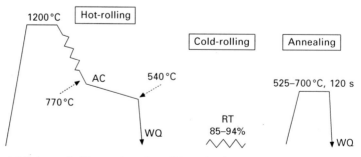

4.1 Schematic illustration describing the fabricating conditions of UFG-FC steel.

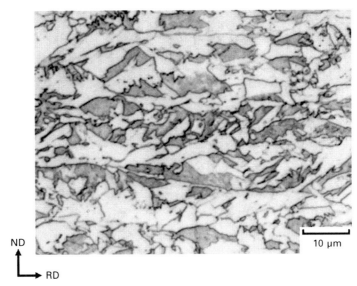

ND

RD

4.2 Optical micrograph of the hot-rolled sheet of the UFG-FC steel. Observed from TD.

Figure 4.3 (a), (d) and (g) shows SEM microstructures of the specimens cold-rolled to reductions of 85%, 91% and 94% in thickness, respectively. The ferrite matrix (dark gray region) exhibited a wavy microstructure elongated roughly to the rolling direction (RD) and bent along the martensite islands (light gray region).

It was indicated that complex plastic flow occurred and higher strain was introduced in the softer ferrite matrix owing to the existence of hard martensite phase. TEM observation[8] of the 91% cold-rolled specimen confirmed a fine lamellar structure with a mean spacing of 0.14 μm in the ferrite region. It was also confirmed by selected area diffraction that this region contained various crystal orientations.[8] It is worth noting that the fine structure including large misorientations, which had been found in SPD-processed materials, already existed at the cold-rolled state in spite of the relatively low plastic strain of 2.8. It is well known that the hard second particles increase the rate of formation of high angle grain boundaries (HAGB) in the matrix through inhomogeneous deformation in two-phase metals and alloys.[22] The large local misorientation is caused by local lattice rotations in the vicinity of the hard particles. In addition, the shear strain should be closely related to the significant grain refinement of ferrite matrix. Kamikawa *et al.*[23] have reported that redundant shear strain introduced at sheet subsurface regions owing to high friction in rolling accelerates the microstructure refinement in the ARB of IF (interstitial-free) steel. The wavy-shaped ferrite and diamond-

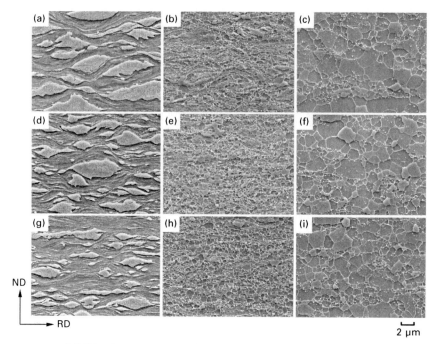

ND ↑

RD →

2 µm

4.3 SEM microstructures of (a), (d), (g), cold-rolled specimen, (b), (e), (h), specimens annealed at 600°C after cold-rolling and (c), (f), (i), specimens annealed at 650°C after cold-rolling. The cold-rolling reductions are (a), (b), (c), 85%, (d), (e), (f), 91% and (g), (h), (i), 94%. Observed from TD.

shaped martensite in Fig. 4.3 suggested that large plastic strains including shear strain components were introduced to the ferrite grains.

On the other hand, the martensite islands were also deformed to some extent in the cold-rolling and showed diamond shapes. Table 4.2 summarises the cold-rolling reduction and the mean thickness ratio, t/t_0, of the martensite islands in the microstructures measured using the OM of the hot-rolled sheet and the SEM micrograph of the cold-rolled specimens.

Here, t and t_0 are the mean intersect lengths of the martensite islands along ND after and before cold-rolling, respectively. The reduction of the martensite islands was much smaller than the reduction of the specimen, which indicated that a larger strain was introduced to the ferrite grains. By TEM analysis in the previous study,[8] large local misorientations in the deformed martensite regions were also confirmed in spite of the smaller strain in martensite. Ueji et al.[24] have reported that 50% cold-rolled low-carbon martensite exhibited fine lamellar structure involving large misorientations, which was equivalent to the microstructure in SPD processed steels. This is thought to be attributed to the complex and fine microstructure of the

Table 4.2 Cold-rolling reduction and mean thickness ratio of the specimens and the martensite islands of the cold-rolled UFG-FC specimens

Cold-rolling reduction of the specimen (%)	t/t_0 of the specimen	Mean t/t_0 of the martensite islands
85	0.15	0.47
91	0.09	0.32
94	0.06	0.27

as-transformed martensite, also involving a high density of dislocations. They also showed equiaxed UFG microstructure after annealing of the 50% cold-rolled specimen with a single phase of martensite. It can be concluded therefore that the strain applied to martensite in the present study was not very large but probably enough to introduce large local misorientations and to form UFG microstructure through subsequent annealing.

Figure 4.3 (b), (e) and (h) show the microstructures of the cold-rolled sheets after annealing at 600 °C. In the 85% cold-rolled and annealed specimen (Fig. 4.3 (b)), both equiaxed fine ferrite grains and elongated ferrite grains located in an arc-like row (in the lower part of Fig. 4.3 (b)) were observed. The fine equiaxed ferrite grains seemed to be formed by continuous coarsening of the finely subdivided regions in the cold-rolled microstructure together with recovery.[8] It was difficult to distinguish clearly which area in Fig. 4.3 (b) was originally ferrite or martensite, but according to the supposed mechanism discussed before, the ultrafine ferrite grains could originate from both ferrite and martensite. On the other hand, some coarse grains in an arc-like row seen were probably formed mainly by recovery of the ferrite regions such as the elongated and bent ferrite seen in the upper and left part of Fig. 4.3 (a). Those ferrite regions in the cold-rolled microstructure seem to be deformed to a smaller plastic strain because of the relatively low density of surrounding martensite islands. Also the fact that the coarse ferrite grains in the annealed microstructure contained few cementite particles suggests that they were originally ferrite. When the cold-rolling reduction was higher than 91%, the specimens were filled mostly with equiaxed ultrafine ferrite grains (Fig. 4.3 (e) and (h)). This seemed to be because the spacing of the martensite islands as seen in Fig. 4.3 (d) and (g) decreased due to larger cold-rolling reductions. When the specimens were annealed at 650 °C, significant grain growth occurred in all the specimens. Fine cementite particles dispersed in ferrite grains were also observed in all annealed specimens in Fig. 4.3. Figure 4.4 shows the relationship between the annealing temperature and mean ferrite grain size measured on the SEM micrographs by the mean intersection method along the RD and ND. The larger the cold-rolling reduction was, the smaller the obtained ferrite grain size became.

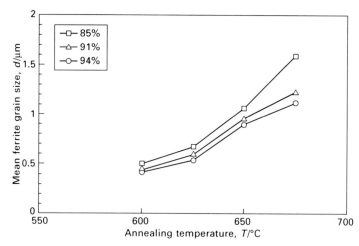

4.4 The relationship of annealing temperatures and mean ferrite grain sizes measured by the mean intersect method of UFG-FC steels.

The sizes of the equiaxed ultrafine ferrite grains shown in Fig. 4.3 (b), (e) and (h) were not so different. However, due to the existence of the elongated and arc-like ferrite grains the average ferrite grain size in the specimen 85% cold-rolled and annealed at 600 °C (Fig. 4.3 (b)) was slightly larger than that in other specimens having larger cold-rolling reductions (Fig. 4.3 (e) and (h)). On the other hand, the effect of the annealing temperature on the ferrite grain size was more significant as shown in Fig. 4.4. The 85% cold-rolled specimens annealed at 600 °C and 625 °C contained some recovered ferrite grains as shown in Fig. 4.3 (b). The minimum grain size of the homogeneous grain structures obtained was 0.43 μm in the specimen 94% cold-rolled and then annealed at 600 °C (Fig. 4.3 (h)).

The UFG ferrite formed throughout the specimens by large cold-rolling reductions followed by low annealing temperatures. In order to clarify the process of fine-grain formation, Fig. 4.5 shows the microstructure of a specimen that was 94% cold-rolled and subsequently annealed at 525 °C for 120 seconds.

The dotted lines roughly delineate the former martensite islands that had already changed to fine-grained ferrite and cementite. In the ferrite matrix, UFG ferrite also formed along the wavy microstructure in the vicinity of the former martensite. Finer equiaxed grains of ferrite were seen around the former martensite islands. The result suggests that a larger strain was introduced into the ferrite/martensite interface regions through cold-rolling. Therefore, in order to reduce the cold-rolling reductions required, it seems to be effective to decrease the spacing of martensite islands in the starting microstructure before cold-rolling.

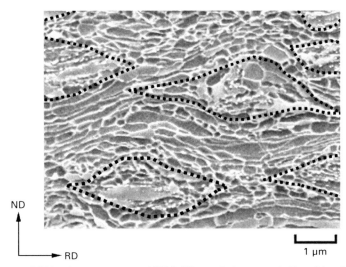

ND

RD   1 µm

4.5 SEM microstructure of UFG-FC steel that was 94% cold-rolled and annealed at 525 °C.

4.3.2   Mechanical properties of nanostructured ferrite-cementite steel sheets

The tensile properties of nanostructured ferrite-cementite steels were investigated using a high-speed servo-hydraulic material test system produced by Saginomiya Inc. equipped with a special load-sensing block.[25,26] With this machine, stress–strain (s–s) curves at a wide range of strain rates from quasi-static to dynamic deformations can be obtained. Figure 4.6 shows the appearance of the prepared tensile specimen having a gauge length of 6mm and a width of 2mm.

The tensile direction was parallel to RD. Tensile tests were operated at various strain rates ranging from $10^{-2}s^{-1}$ to $10^{3}s^{-1}$ at room temperature. Total elongation of the specimens was measured from the difference in the gauge length before and after testing. Fabricating conditions, microstructures and quasi-static mechanical properties measured at a strain rate of $10^{-2}s^{-1}$ are summarised in Table 4.3.

UFG-FC A, B, C and FCM specimens were prepared by 91% cold-rolling of the same hot-rolled sheet as shown in Fig. 4.2 and subsequent annealing at 620 °C, 635 °C, 670 °C and 700 °C for 120 seconds, respectively. The mean ferrite grain sizes indicated in Table 4.3 were calculated by the EBSD (electron backscatter diffraction) data measured on the TD sections. UFG-FC steels showed microstructures of ultrafine ferrite and cementite, while the FCM specimen showed a microstructure composed of ferrite, cementite and martensite. FCM specimen contained about 14% of martensite in the

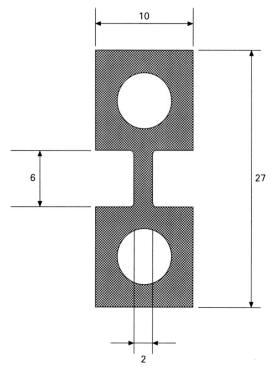

4.6 Schematic drawing of the test piece for dynamic and quasi-static tensile tests.

microstructure. The A_{c1} transformation temperatures of UFG-FC steel measured by a dilatometer was approximately 700 °C, so that the FCM specimen contained both ferrite and austenite during the annealing at 700 °C and the austenite transformed to martensite during the subsequent water-cooling.

Figure 4.7 shows s–s curves at strain rates of 10^{-2}, 10^2 and 10^3 s^{-1}. Figure 4.7 (a), (b), (c) and (d) correspond to the test results of UFG-FC steels A, B, C and the FCM steel listed in Table 4.3.

Both UFG-FC and FCM steels showed the yield drop phenomenon. The flow stress increased whereas the uniform elongation significantly decreased as the ferrite grain size in the UFG-FC specimens decreased (Fig. 4.7 (a), (b) and (c)). The same behavior has been reported for UFG IF steel sheets prepared by ARB and the annealing process.[18] On the other hand, as shown in Fig. 4.7 (d), the work hardening increased when martensite was introduced in the microstructure.

The flow stress significantly increased when the strain rate increased in UFG-FC steels. In order to investigate the strain rate dependence of the flow stress, the difference in tensile flow stress at 5% nominal strain between the strain rates of 10^3 s^{-1} and 10^{-2} s^{-1}, $\Delta\sigma$, was evaluated. Figure 4.8 shows the

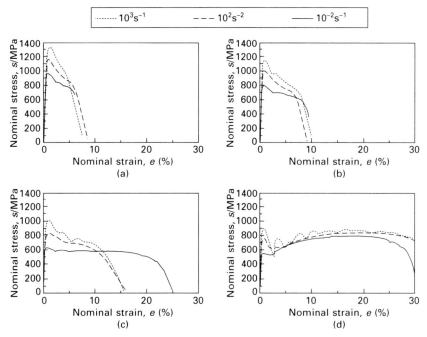

4.7 Nominal stress-strain curves of UFG-FC steel specimens 91% cold-rolled and annealed at (a) 620°C, (b) 635°C, (c) 670°C and (d) 700°C.

Table 4.3 Microstructures and quasi-static tensile properties of the fabricated UFG steels and FCM steel for tensile tests

	Annealing temperature (°C)	Microstructure	Mean ferrite grain size (μm)	0.2% offset stress (MPa)	Tensile strength (MPa)	Total elongation (%)
UFG-FC A	620	F-C	0.49	966	966	8.4
UFG-FC B	635	F-C	0.62	816	820	11.3
UFG-FC C	670	F-C	0.95	636	638	23.4
FCM	700	F-C-M	1.0	515	753	28.0

Note: F: ferrite, C: cementite, M: martensite.

relationship between $\Delta\sigma$ and the quasi-static flow stress of the fabricated UFG-FC steels and other conventional steels referred from previous studies (some were evaluated at slightly different strains).

The present UFG-FC includes additional data not listed in Table 4.3. The $\Delta\sigma$ of UFG-FC steels with sub-micron grain sizes were as large as that of mild steels[27,28] with a single ferrite phase. However, the conventional high-strength steels (HSSs)[27–29] showed a lower $\Delta\sigma$. The FCM specimen fabricated in this study had a similar $\Delta\sigma$ value to conventional HSSs. In

Fig. 4.8 the data of UFG-FC steels reported by Tsuchida *et al.*[21] are plotted as well. The $\Delta\sigma$ values by Tsuchida *et al.* were close to that of the present work. This result indicates that the grain refinement of FC microstructure to sub-micron size did not decrease the $\Delta\sigma$. It is expected that the large $\Delta\sigma$ value promotes high dynamic energy absorption due to the large dynamic flow stress.

Generally in most metals and alloys the flow stress increases with an increasing strain rate.[30-33] This is explained in terms of the thermally activated process of the dislocation motion against short-range obstacles such as Peierls potential. The higher the strain rate, the higher the external stress required for the dislocation motion. During the deformation, the thermal vibration of iron atoms can help the dislocations to overcome the obstacles. As the strain rate is increased, the assist by the thermal vibration will be less effective because there is less time available to overcome the obstacles. Therefore, at higher strain rates, the required external stress for dislocation motion will be increased. The component of flow stress that is related to the thermal activation process is called thermal stress, while the other component, which is independent of the strain rate and temperature, is called athermal stress. When alloy elements are added, the strain rate sensitivity decreases.[33] Among commercial steels, mild steels show the highest strain rate sensitivity in flow stress and it decreases gradually when the steels are strengthened by alloy addition (solution hardening) or the introduction of second phases (precipitation hardening).[27,28,34,35] This is understood in terms of the increase of the activation volume in the thermally activated process of dislocation motion owing to alloy elements, precipitates and second phases.[36] On the

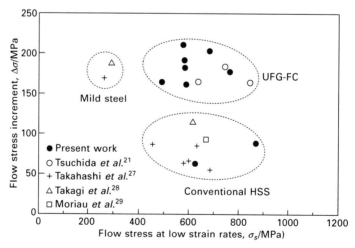

4.8 Relationship of the difference in flow stress, $\Delta\sigma$ and quasi-static flow stress of fabricated UFG-FC steels and other steels in references.

other hand, it has been shown that the UFG-FC microstructure does not decrease the strain rate sensitivity in flow stress.[9,20,21] It means that ultra grain refinement influenced mainly the athermal component of flow stress rather than the thermal component.

4.3.3   Further requirements for nanostructured ferrite-cementite steel sheets

In the previous section, it was shown that ultra grain refinement increased strength significantly while keeping a large strain rate sensitivity in flow stress. The strain rates of the material through the dynamic axial collapse of square columns exceed 10^2 s^{-1},[37] therefore the dynamic strength of the material is important for automotive body parts in the crushable zones. Dynamic collapse tests of hat columns confirmed good performance of fabricated UFG-FC steel sheets. However, ultra grain refinement reduced the uniform elongation of the steel (Fig. 4.7). This is due to the insufficient work hardening rate in UFG single-phase microstructures.[38,39] Based on the plastic instability condition in tensile tests, the highly strengthened material requires a greater work hardening rate to keep the uniform deformation. The work hardening rate of the present UFG-FC steels seemed insufficient, resulting in early necking (early plastic instability) in the tensile test. Considering the stamping process of the automotive body parts, adequate uniform elongation is required. Therefore in order to apply UFG steels widely to automotive body parts, the uniform elongation should be improved. In order to meet this demand, the best solution seems to be an introduction of hard second phases to the UFG microstructure.[40] Generally duplex microstructures, for example dual-phase (DP) or multi-phase (MP) microstructures, show improved uniform elongation by the enhanced work hardening rate. Also in the case of fine-grained steels, a calculation using the 'secant method'[41] has shown that significant improvement could be expected by an introduction of a hard phase. In the next section, the concept of improving the uniform elongation in UFG steels is shown, and the test results using fabricated UFG-MP steels are described and discussed.

4.4   Improving elongation in nanostructured steel sheets

4.4.1   Introduction of hard second phases to nanostructured ferritic microstructures

In order to improve the uniform elongation of nanostructured or UFG ferritic steels, one of the potential ways is to introduce hard phases.[40] A recent study on UFG-DP steels using ECAP and subsequent intercritical annealing[42] has shown that the UFG-DP microstructure with a mean ferrite grain size of 0.8 to 1.2 μm improved tensile strength without loss of uniform elongation compared with a coarse grained DP steel. In the present work, as shown in Table 4.3, introducing martensite by increasing the final annealing temperature improved both the tensile strength and elongation. However, the obtained tensile strength was not so high because the ferrite grain size was around 1 μm, owing to the grain growth. The potential competitor of UFG steels is modern UHSSs, which have higher tensile strengths than 1000 MPa.[1,13]

In this section, we show the concept of introducing hard second phases while keeping the ultrafine ferrite matrix. In order to avoid grain growth, the transformation temperature of the steel was decreased by Mn addition, which could decrease the intercritical annealing temperatures. Table 4.4 shows the chemical compositions of the steels used in this section.

The A_{c1} transformation temperatures of UFG-MP1 and UFG-MP2 steels measured by a dilatometer were 690 °C and 650 °C, respectively. The transformation temperatures were significantly lower than that of UFG-FC steel (approximately 700 °C in UFG-FC steel). The fabricating process of UFG-MP1 and UFG-MP2 steels are shown in Fig. 4.9 (a) and (b), respectively. For UFG-MP1 steel, after hot-rolling to a thickness of 6 mm, the sheet was immediately cooled to room temperature by water, followed by intercritical annealing at 730 °C for 90 minutes in order to obtain a duplex microstructure of ferrite and martensite. After mechanical grinding, the hot-rolled sheets of UFG-MP1 steel were cold-rolled from 5 mm to 1.5 mm in thickness at room temperature. The total reduction through cold rolling was 70%. The cold-rolled sheets were annealed for 120 seconds at 675 °C or 700 °C followed by water-cooling to room temperature. The annealing temperature of 700 °C corresponded to the intercritical region of ferrite and austenite. On the other hand, as shown in Fig. 4.9 (b), UFG-MP2 steel was air-cooled to room temperature after hot-rolling to a thickness of 5 mm. The air-cooling rate was sufficient to obtain a complex structure composed of soft ferrite and hard phases in UFG-MP2 steel due to the higher Mn content. The hot-rolled sheets of UFG-MP2 steel were heated at 500 °C for 1 hour in order to reduce the hardness, and then rolled from 3.5 mm to 1.2 mm in thickness without reheating. The total reduction through the warm-rolling was 66%. The warm-rolled sheets were annealed for 120 seconds at intercritical temperatures of 685 °C or 700 °C, followed by subsequent annealing at 400 °C for 300 seconds. The austenite amounts of the specimens after the final annealing

Table 4.4 Chemical composition (mass%) of UFG-MP steels

Steel	C	Si	Mn	P	S	Al	Nb	N
UFG-MP1	0.21	0.01	3.0	0.002	0.001	0.022	0.046	0.0013
UFG-MP2	0.15	1.42	4.0	0.001	0.001	0.026	0.054	0.0021

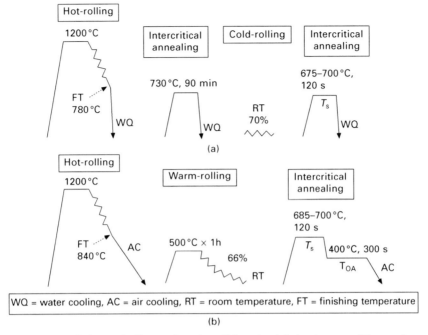

4.9 Schematic illustration describing the fabricating conditions of (a) UFG-MP1 steel and (b) UFG-MP2 steel.

were estimated by X-ray diffractometry using Co-Kα radiation. The austenite volume fraction was calculated using the integrated intensities of (110)α, (200)α, (211)α, (111)γ, (200)γ and (220)γ diffraction peaks.

Figure 4.10 (a) shows the OM after the intercritical annealing of the hot-rolled sheet of UFG-MP1 steel.

A fine duplex microstructure of ferrite (light gray regions) and martensite (dark gray regions) was observed. The mean intersect lengths of each phase in the hot-rolled microstructure, t_0, were measured using the OM of the hot-rolled sheet shown in Fig. 4.10 (a). As summarised in Table 4.5, the t_0 values along ND of the ferrite and martensite regions were 1.65 μm and 1.73 μm, respectively. The area of ferrite region in the hot-rolled microstructure measured by point counting method was 30%. Figure 4.10 (b) shows the SEM microstructure of the specimen after cold-rolling. It was shown that the ferrite grains (dark regions) were elongated and had wavy shapes, which meant that large and complex plastic deformation was introduced in the ferrite. The martensite (gray regions) was also deformed roughly along the RD. The features are the same as those in the cold-rolled microstructure of UFG-FC steels shown in Fig. 4.3.

Table 4.5 also summarises the mean intersect lengths after cold-rolling (t) and the mean thickness ratios (t/t_0) of both the ferrite and martensite regions.

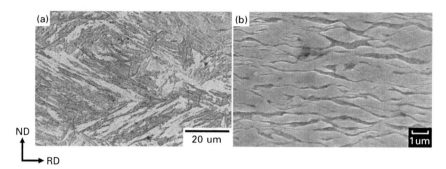

ND
RD

4.10 (a) Optical micrograph of the hot-rolled and annealed sheet (before the cold-rolling) of UFG-MP1 steel observed from TD. (b) SEM microstructure of the cold-rolled sheet of UFG-MP1 steel observed from TD.

Table 4.5 Mean intersect lengths before the cold-rolling (t_0), after the cold-rolling (t) and the mean thickness ratios (t/t_0) of both ferrite and martensite regions in UFG-MP1 steel

Phase	Mean intersect length before cold-rolling, t_0	Mean intersect length after cold-rolling, t	Mean thickness ratio, t/t_0
Ferrite	1.65	0.18	0.11
Martensite	1.73	0.52	0.30

The t values were measured using the SEM micrograph of the cold-rolled specimen shown in Fig. 4.10 (b). The t/t_0 of the martensite regions was approximately 0.3, which was close to the t/t_0 of the specimen. On the other hand, the t/t_0 value of 0.11 in the ferrite regions was much smaller than that of the specimen. This result indicates that the plastic strain concentrated to the ferrite region through cold rolling.

Figure 4.11 shows SEM microstructures after the final annealing of the cold/warm-rolled sheets of UFG-MP1 (Fig. 4.11 (a) and (b)) and UFG-MP2 (Fig. 4.11 (c) and (d)). The intercritical annealing temperatures of the specimens in Fig. 4.11 (a), (b), (c) and (d) are 675 °C, 700 °C, 685 °C and 700 °C, respectively. In both steels, the volume fraction of the second phases (the light gray regions in Fig. 4.11) increased with the annealing temperature.

Here, 'second phases' means austenite, bainite and martensite which appear as light gray islands in the SEM micrographs. Table 4.6 summarises the ferrite grain size, the area share of the second phases in the SEM micrograph and the retained austenite content measured by X-ray diffraction. The rest of the second phases should be bainite and/or martensite as observed in low-C TRIP (transformation-induced plasticity) steels;[43,44] however, those phases were not distinguished clearly in the SEM micrograph. It was shown that a

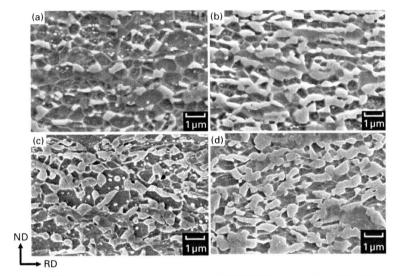

4.11 SEM microstructures of fabricated UFG-MP steels. (a) UFG-MP1 steel annealed at 675°C (b) UFG-MP1 steel annealed at 700°C, (c) UFG-MP2 steel annealed at 685°C and (d) UFG-MP2 steel annealed at 700°C. Observed from TD.

Table 4.6 Annealing temperature and microstructural parameters of fabricated UFG-MP steels

Steel	Annealing temperature (°C)	Mean ferrite grain size (μm)	Fraction of second phases (%)	Austenite fraction (%)
UFG-MP1	675	0.58	17	13
UFG-MP1	700	0.51	36	16
UFG-MP2	685	0.49	44	12
UFG-MP2	700	0.41	50	22

considerable amount of hard second phases were introduced by keeping the ferrite grain size less than 1 μm, owing to the decreased A_{c1} transformation temperatures in UFG-MP steels with relatively high Mn content. Also it is worth noting that UFG microstructures were obtained in spite of the relatively lower cold-rolling reductions (70% in the UFG-MP1 steel). As summarised in Table 4.5, in UFG-MP1 steel the mean intersect length of the ferrite region before the cold-rolling was 1.65 mm, which was much smaller than that of UFG-FC steel (5.55 μm). It seemed that the reduction of 70% was sufficient to concentrate the plastic strain to the ferrite regions due to the small intersect length. This result indicates that the grain refinement of the starting microstructure is effective in reducing the required cold-rolling reduction for the formation of UFG microstructures through subsequent annealing.

4.4.2   Mechanical properties of nanostructured multi-phase steel sheets

Tensile properties were investigated at a nominal strain rate of $10^{-2}\,s^{-1}$ at room temperature using the specimens shown in Fig. 4.6. The tensile direction was parallel to RD. Figure 4.12 shows s–s curves of fabricated UFG-MP steels with various annealing temperatures. Yield drop with Lüders elongation was observed in all the specimens.

However, significant work hardening was observed after the Lüders elongation, except for UFG-MP1 steel annealed at 675 °C, which resulted in a good combination of high strength and large uniform elongation. The mechanical properties are summarised in Table 4.7.

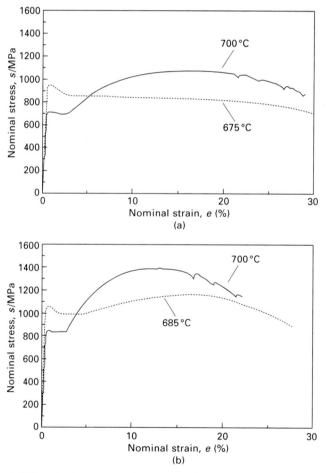

4.12 Nominal stress-strain curves of (a) the UFG-MP1 steel and (b) the UFG-MP2 steel corresponding to Fig 4.11. The temperatures in the figure indicate the intercritical annealing temperatures.

Table 4.7 Annealing temperature, thickness and quasi-static mechanical properties of fabricated UFG-MP steels

Steel	Annealing temperature (°C)	Thickness (mm)	0.2% offset stress	Tensile strength (MPa)	n-value (5–10%)	Uniform elongation (%)	Total elongation (%)
UFG-MP1	675	1.5	937	937	0.056	0.6	32
UFG-MP1	700	1.5	719	1050	0.324	13	25
UFG-MP2	685	1.2	1034	1155	0.205	17	28
UFG-MP2	700	1.2	840	1380	0.321	13	22

Here, the *n*-values, which represent the work hardening rate of the steels, were calculated assuming that s–s curves at a strain range of 0.05 to 0.1 could be approximated by:

$$\sigma = K\varepsilon^n,$$ [4.1]

where σ, K and n are true stress, a constant, and a work hardening component (*n*-value), respectively.

The *n*-values of the fabricated steels were calculated by:

$$n = \ln\left(\frac{\sigma_2}{\sigma_1}\right)\Big/\ln\left(\frac{\varepsilon_2}{\varepsilon_1}\right),$$ [4.2]

where σ_2 and σ_1 are true stress at plastic strains of ε_2 and ε_1, respectively. In this case plastic tensile strains (true strain) of 0.05 and 0.1 were chosen. The UFG-MP steels, except for UFG-MP1 steel annealed at 675 °C, had fairly larger *n*-values than recent UHSSs.[1,45,46]

The significant work hardening of UFG-MP steels seemed to be caused by strain-induced martensitic transformation of the retained austenite, since the austenite fraction after 15% tensile strain was less than 1% in UFG-MP1 steel annealed at 700 °C. In fact, it has been reported that low-carbon steels containing 5% to 7% Mn show good a combination of strength and elongation[47–51]. In Merwin[48], a multi-phase microstructure of 7% Mn steel composed of fine ferrite and hard phases has been shown. The concept of this microstructure is close to that of UFG-MP steels in the present study. However, the strength–elongation balance in the present UFG-MP steel containing 4% Mn was better than that of 5% Mn steels in the previous studies.[47–50] The difference seems to be caused by the difference of the annealing conditions. In the present annealing process shown in Fig. 4.9 (b), the heating rate was high (10 °C s⁻¹ to 25 °C s⁻¹) and the holding time at the intercritical region was short (120 seconds), due to heating with salt bath. On the other hand, in the previous studies[47–49] 'batch-annealing' was employed. For example, the heating rate was 0.16 °C s⁻¹ and the holding time was 1200 seconds.[50] The rapid heating and short holding time in the present study were effective to suppress the grain growth, which seemed to improve the combination of strength and elongation.

In addition, the dynamic tensile tests were carried out also for UFG-MP steels using the same equipment described in Section 4.3.2. The $\Delta\sigma$ value (the difference in tensile flow stress at 5% nominal strain between the strain rates of 10^3 s^{-1} and 10^{-2} s^{-1}) was around 100 MPa in UFG-MP1 steel. This value was close to that in conventional HSSs and not so large compared with that in UFG-FC steels. In the UFG microstructure, the strain rate sensitivity in flow stress decreased by the introduction of hard second phases as well as in the conventional coarse-grained microstructure. This result suggests that the thermal activation process of dislocation motion was controlled mainly by the hard second phases as described in Section 4.3.2, even in the UFG microstructure.

4.5   Crash-worthiness of nanostructured steel sheets

In order to evaluate the crash-worthiness for automotive body parts, dynamic collapse tests[46] have been widely employed. In this study, dynamic axial collapse tests by using hat rectangular columns were carried out. Figure 4.13 shows the cross-section of the hat column, which is composed of a hat-shape part and a flat back plate. The hat-shape part, which was prepared by bending steel sheet with a radius of 5 mm, was set with a back plate and then spot-welded at the center of the flanges. The same material was used for both the hat part and the back plate. The pitch of the spot-welding was 40 mm.

Figure 4.14 shows a schematic illustration of the dynamic collapse test. A free-fall type collapse test machine was employed. A 110 kg weight was dropped from 11 m and hit the top end of the hat column fixed vertically on a steel plate by arc-welding. The weight displacement was measured by a laser-type displacement sensor, and the deformation load was measured

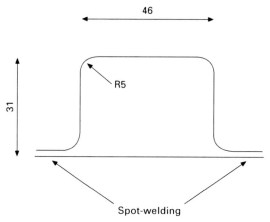

4.13 Schematic drawing of the cross-section of the hat column.

4.14 Appearance of the hat column for the dynamic collapse test.

by a load-cell set under the hat column. Two steel pipes were set beside the hat column in order to stop the falling weight at a collapse displacement of 160 mm.

Two types of UFG-MP steels, UFG-MP1 and the UFG-MP2, were evaluated in the dynamic collapse tests. The chemical compositions of both steels were the same as those shown in Table 4.4 The process was basically the same as that shown in Fig. 4.9. The fabricating conditions of the steels are shown in Table 4.8.

The FT, T_s and T_{OA} in Table 4.8 mean the temperature after hot-rolling, the temperature of intercritical annealing and the holding temperature after intercritical annealing, respectively. The CR and WR mean cold-rolling and warm-rolling, respectively. In warm-rolling, the hot-rolled sheets were heated at 500 °C for 1 hour and then rolled to 1.2 mm in thickness without reheating. They are indicated in Fig. 4.9. In UFG-MP1 steel, the intercritical

Table 4.8 Fabricating conditions of UFG-MP steels used for axial collapse tests. The meanings of FT, T_S, and T_{OA} are the same as in Fig. 4.9

Steel	FT (°C)	Coolingq after HR	CR/WR	Thickness of sheet (mm)	CR/WR reduction (%)	T_s (°C)	T_{OA} (°C)
UFG-MP1	780	AC	CR	1.2	76	700	–
UFG-MP2	840	AC	WR	1.2	66	690	400

Note: HR: hot-rolling, WQ: water-cooling, AC: air-cooling, CR: cold-rolling, WR: warm-rolling.

annealing after hot-rolling was replaced by direct air-cooling, and the thickness of the specimen after cold-rolling was reduced to 1.2 mm. The UFG-MP2 specimens were prepared in almost the same conditions as those shown in Fig. 4.9 (b) except for the intercritical annealing temperature. For comparison, a commercially available DP steel sheet was also tested. Figure 4.15 shows quasi-static s–s curves of the evaluated steels obtained using JIS No. 5 specimens with a gauge length of 50 mm and a gauge width of 25 mm, of which the direction was parallel to the RD. The UFG-MP1 and MP2 steels showed steady work hardening up to tensile strains of 16–17%, while it saturated at the early stage of tensile deformation in the DP steel.

Table 4.9 summarises the mechanical properties of the steels. All steels had a tensile strength higher than 1000 MPa; however, the *n*-values in UFG-MP steels were significantly larger than those of DP steel.

Figure 4.16 shows the appearance of deformed columns following dynamic collapse tests. The columns of UFG-MP steel were deformed in a compact-mode,[52,53] which means a stable deformation through the formation of a series of continuous folds like an accordion.

This is the ideal deformation mode in the axial collapse for impact energy absorption. On the other hand, the column made of DP steel did not form compact buckling but fell down due to the moment of rotation generated by the non-deformed straight region, and buckling near the bottom end of the column. Also fracture at the hat wall was observed. This is called 'non-compact'[52,53] mode, which means an unstable deformation through the formation-retaining straight regions in the column walls. The change in deformation mode from compact to non-compact, which reduces the efficiency of energy absorption, is caused by the decreased work hardening rate of the material highly strengthened.[16,17] Commercially available UHSS generally has a low work hardening rate (*n*-value)[45,46] because it has more hard phases such as martensite in the microstructure. Uenishi *et al.*[17] showed by FEM (finite element model) simulation that a desirable compact-mode deformation of hat columns during axial collapse could be promoted by improving the work hardening rate of the material. The authors[10] confirmed by FEM simulation that the high *n*-value promoted continuous strain distribution between buckling regions, which resulted in a continuous folding (compact

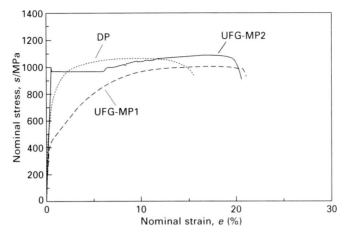

4.15 Quasi-static s-s curves of the steels evaluated in the dynamic collapse tests. The dashed line, the solid line and the dotted line represent the UFG-MP1, UFG-MP2 and DP steels. JIS No.5 specimens were used.

Table 4.9 Quasi-static tensile properties of UFG-MP steels and DP steel used for axial collapse tests. JIS No. 5 specimens were used

Steel	Thickness (mm)	0.2% offset stress (MPa)	Tensile strength (MPa)	n-value (5–10%)	Uniform elongation (%)	Total elongation (%)
UFG-MP1	1.2	453	1005	0.244	17	22
UFG-MP2	1.2	972	1088	0.284	16	21
DP	1.2	747	1064	0.068	8	14

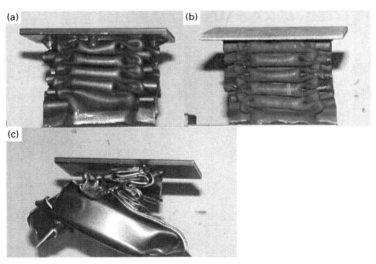

4.16 Photographs of the hat columns after dynamic collapse tests. (a) UFG-MP1 steel, (b) UFG-MP2 steel and (c) DP steel.

mode), while strain concentration in the buckling areas was significant and resulted in non-compact folding in a material with low *n*-value. It was shown that the improved work hardening rate promoted the compact mode in the axial collapse.

Figure 4.17 (a) shows the load-displacement curves of the two columns, made of UFG-MP2 steel and DP steel.

In the case of these thin wall columns, the maximum load corresponds to the first peak load at the start of buckling. In both steels the deformation loads were almost the same until the collapse displacement of 100 mm; however, the load dropped in DP steel after displacement of about 100 mm. On the other hand, the deformation load of the UFG-MP2 column kept the same level until displacement of 160 mm. It was confirmed by the high-speed camera observation that the load drop in the DP column corresponded

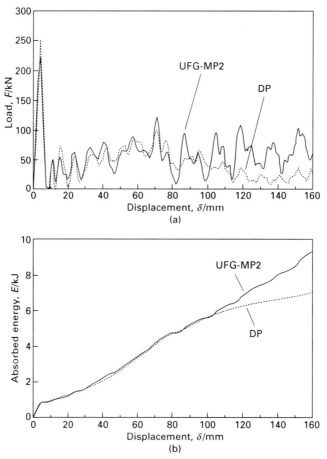

4.17 (a) Load-displacement curves and (b) absorbed energy-displacement curves obtained by dynamic collapse tests.

to the onset of the decline of the column. As a result, as shown in Fig. 4.17 (b), the absorbed energy saturated at the displacement of 100 mm in the DP steel. On the other hand, in the column of UFG-MP2 steel the absorbed energy increased continuously until displacement of 160 mm. This continuous energy absorption increment of UFG-MP2 steel is favourable for the front and rear frames in automotive body structures that are deformed along the longitudinal direction in a collision. Table 4.10 summarises the results of the dynamic collapse tests.

The maximum load of the DP column was the largest due to its high yield strength. However, the average load and absorbed energy in UFG-MP steels were fairly larger than those in DP steel. This is because of the steady deformation load brought about by an ideal compact-mode deformation in UFG-MP steels. The ratio of the average deformation load to the maxim load was greater in UFG-MP steels, which was also favorable in the axial collapse. The UFG-MP1 specimen for the collapse test exhibited Lüders elongation. The Lüders elongation was capable of causing irregularities such as local buckling or strain concentration in the axial collapse of the columns. However, the effect of the Lüders elongation on the deformation behaviour was not observed clearly. It could be concluded that work hardening at a tensile strain over 5% was important for promoting the compact mode. Furthermore, if new microstructures are obtained with both a large work hardening rate and a large strain rate sensitivity, they should be promising materials for automotive body structures.

4.6   Conclusions

In this chapter the future demands for nanostructured or ultrafine-grained (UFG) steel sheets for automotive body structures were described first, and practical test results for evaluating UFG steel sheets were presented. A new route to fabricating UFG steel sheets without severe plastic deformation was introduced. It was shown that UFG ferrite-cementite (FC) steel showed large strain rate sensitivity compared with other high-strength steels (HSSs), while uniform elongation was significantly decreased by ultra grain refinement. In order to improve the uniform elongation of the UFG ferritic microstructure, UFG multi-phase (MP) steels were developed. The UFG-MP steel sheets showed an improved strength–elongation balance. The large work hardening

Table 4.10 Axial collapse properties of UFG-MP steels and DP steel

Steel	Maximum load (kN)	Average load (10–150 mm) (kN)	Average load/ maximum load	Absorbed energy (0–150 mm) (kJ)
UFG-MP1	219	49.1	0.22	7.73
UFG-MP2	223	55.1	0.25	8.60
DP	250	42.0	0.17	6.86

rate of UFG-MP steels promoted a stable compact-mode deformation in dynamic axial collapse of a hat column, which should contribute to the improvement of the crash-worthiness and further weight reduction of automotive body structures.

4.7   References

1   Takechi H. *JOM* 2008;60 No.12: 22.
2   Matlock D.K., Speer J.G. Third generation of AHSS: microstructure design concepts. In: Haldar A., Suwas S., Bhattacharjee D., editors. *Microstructure and Texture in Steels and Other Materials*. London: Springer, 2009; p. 185.
3   Segal V.M. *Mater Sci Eng A* 1995:197: 157.
4   Valiev R.Z., Korznikov A.V., Mulyukov R.R. *Mater Sci Eng A* 1993:168: 141.
5   Saito Y., Tsuji N., Utsunomiya H., Sakai T., Hong R.G. *Scripta Mater* 1998:39: 1221.
6   Tsuji N., Saito Y., Utsunomiya H., Tanigawa S. *Scripta Mater* 1999:40: 795.
7   Saito Y., Utsunomiya H., Tsuji N., Sakai T. *Acta Mater* 1999:47: 579.
8   Okitsu Y., Takata N., Tsuji N. *Scripta Mater* 2009:60: 76.
9   Okitsu Y., Takata N., Tsuji N. *J Mater Sci* 2008:43: 7391.
10 Okitsu Y., Naito T., Takaki N., Sugiura T., Tsuji N. SAE Technical Paper 2010: 2010-01-0438.
11 Kuriyama Y., Takahashi M., Ohashi H. *J Soc Automotive Eng Jpn* 2001:55 No.4: 51.
12 Lüdke B., Pfestorf M. Functional design of a "Lightweight body in white" – How to determine body in white materials according to structural requirements. In: Hashimoto S., Jansto S., Mohrbacher H., Siciliano F., editors. *International Symposium on Niobium Microalloyed Sheet Steel for Automotive Applications*. Warrendale: TMS, 2006; p. 27.
13 World Steel Association, 2009. Advanced High Strength Steel (AHSS) Application Guidelines Version 4.1. Available from: http://www.worldautosteel.org [accessed 15 January 2010].
14 Grassel O., Krüger L., Frommeyer G., and Meyer L.W. *Int J Plast* 2001:16: 1391.
15 Cornette D., Cugy P., Hildenbrand A., Bouzekri M., Lovato G. SAE Technical Paper 2005: 2005-01-1327.
16 Mallen R.Z., Odell J., O'Hara B. SAE Technical Paper 2009: 2009-01-0088.
17 Uenishi A., Kuriyama Y., Takahashi M. *Nippon Steel Tech Rep* 2000:81: 17.
18 Tsuji N., Ito Y., Saito Y., Minamino Y. *Scripta Mater* 2002:47: 893.
19 Tsuji N., Okuno S., Koizum Y., Minamino Y. *Mater Trans* 2004:45: 2272.
20 Jia D, Ramesh KT, Ma E. *Acta Mater* 2003:51: 3495.
21 Tsuchida N., Masuda H., Harada Y., Fukaura K., Tomota Y., Nagai K. *Mater Sci Eng A* 2007:488: 446.
22 Humphreys F.J., Hatherly M., *Recrystallization and Related Annealing Phenomena*, 2nd edn. Oxford: Elsevier, 2004. p. 457.
23 Kamikawa N., Sakai T., Tsuji N. *Acta Mater* 2007:55: 5873.
24 Ueji R., Tsuji N., Minamino Y., Koizumi Y. *Acta Mater* 2002:50: 4177.
25 Tanimura S., Mimura K., Umeda T. *Journal de Phys IV* 2003:110: 385.
26 Chuman Y., Kimura K., Tanimura S. *Int J Impact Eng* 1997:19: 165.

27 Takahashi M., Uenishi A., Yoshida H., Kuriyama Y. SAE Technical Paper 2003: 2003-01-2765.
28 Takagi S., Tokita Y., Sato K., Shimizu T., Hashiguchi K., Ogawa K., Mimura K., Tanimura S. SAE Technical Paper 2005: 2005-01-0494.
29 Moriau O., Tosal-Martinez L., Verieysen P., Degrieck J *Int Conf on TRIP-Aided High Strength Ferrous Alloys*, Aachen, 2002; p. 247.
30 Harding J. *Acta Metall* 1969:17: 949.
31 Campbell J.D., Ferguson W.G. *Phil Mag* 1970:21: 63.
32 Harding J. *Mater Tech* 1977:4: 6.
33 Harding J. The effect of high strain rate on material properties. In: Blazynski, T.Z., editor. *Materials at High Strain Rates*. London: Elsevier, 1987; p. 133.
34 Miura K., Takagi S., Hira T., Furukimi O., Tanimura S. SAE Technical Paper 1998: 980952.
35 Bruce D.M., Matlock D.K., Speer J.G., De A.K. SAE Technical Paper 2004: 2004-01-0507.
36 Kato M. *Introduction to the Theory of Dislocations*. Tokyo: Shokabo, 1999; p. 102.
37 Otubushin A. *Int J Impact Eng* 1998:21: 349.
38 Ma E. *Acta Mater* 2004:52: 1699.
39 Tsuji N. Fabrication of bulk nanostructured materials by accumulative roll bonding (ARB). In: Zehetbauer M.J., Zhu Y.T., editors. *Bulk Nanostructured Materials*. Weinheim: Wiley-VCH, 2009; p. 246.
40 Tomota Y., Narui A., Tsuchida N. *ISIJ Int* 2008:48: 1107.
41 Tsuchida N., Tomota Y., Nagai K. *Tetsu-to-Hagané* 2003:89: 1170.
42 Son Y.I., Lee Y.K., Park K.T., Lee C.S., Shin D.H. *Acta Mater* 2005:53: 3125.
43 Jacques P.J., Ladrière J., Delannay F. *Metall Mater Trans A* 2001:32A: 2759.
44 Zaefferer S., Ohlert J., Bleck W. *Acta Mater* 2004:52: 2765.
45 Takakura K., Takagi K., Yoshinaga N. SAE Technical Paper 2006: 2006-01-1586.
46 Walp M.S. SAE Technical Paper 2007: 2007-01-0342.
47 Miller RL. *Met. Trans* 1972:3: 905.
48 Merwin M.J. SAE Technical Paper 2007: 2007-01-0336.
49 Merwin M.J. *Iron & Steel Tech* 2008 Oct: 66.
50 Huang H., Matsumura O., Furukawa T. *Mat Sci Tech* 1994:10: 621.
51 Furukawa T., Huang H., Matsumura O. *Mat Sci Tech* 1994:10: 964.
52 Mahmood H.F., Paluszny A. SAE Technical Paper 1981: 811302.
53 Reid S.R., Reddy T.Y., Gray M.D. *Int J Mech Sci* 1986:28: 295.

4.8   Appendix

Lightweight automotive body structures by steels have been proposed through several projects. The recent projects are reviewed below.

ULSAB (Ultralight steel auto body)

This is the first organized lightweight steel body project from 1994 to 1998. It was operated through a consortium composed of 35 steel manufacturers representing 18 countries. Targeting mid-sized four-door sedans, a practical concept body was constructed. Various kinds of HSS were applied to more than 90% of the body structure. Also several new manufacturing methods, i.e. tailored blanks (steel sheets assembled from a number of single sheets having various thickness and strength by laser welding), tubular hydroforming (forming of a steel pipe by hydro pressure applied to the inside of the tube) and laser-welding assembly were applied. A weight reduction of 25% compared with the benchmark cars was achieved. At the same time, two other projects, ULSAC (Ultralight steel auto closure) and ULSAS (Ultralight steel auto suspension) were carried out.

ULSAB-AVC (ULSAB-advanced vehicle concepts)

As a post-project of ULSAB, advanced lightweight steel bodies, which adapt to more variations of automobile safety assessments, were studied. AHSSs (advanced high strength steels) including DP (dual-phase) and TRIP (transformation induced plasticity) -assisted steels were applied. Targeting European C-class cars and North American midsize-class cars, lightweight car concepts including the body structure, closure panels, chassis and power trains were proposed.

FSV (future steel vehicle)

This project started in 2008 targeting potential cars for mass production in 2020, i.e. BEV (battery electric vehicle), PHEV (plug-in hybrid electric vehicle) and FCEV (fuel cell hybrid electric vehicle). In those vehicles, the main body structure might be different from the conventional ones, i.e. a new structure supporting new power plants. Therefore significant weight reduction could be expected from the new concepts of structures and materials investigated through this project.

Further information is available at the website of World Steel Association: http:// www.worldautosteel.org/.

Aluminium sheet for automotive applications

M. BLOECK, Novelis Switzerland SA, Switzerland

Abstract: This chapter gives an overview on the aluminium sheet alloys that are currently used for outer, inner and structural automotive applications. This includes the typical mechanical properties, such as strength and formability, and information on the corrosion resistance of the various alloys. Furthermore, the different surface modifications are described that are currently applied on the aluminium strip surface, such as special surface topographies, chemical and electrochemical treatments, lubricants and primer coatings. The advantages of these surface modifications during vehicle manufacturing are discussed.

Key words: aluminium auto sheet alloys, auto sheet properties, chemical and electrochemical strip treatments, Aluminium Vehicle Technology system, structural bonding.

5.1   Introduction

An overview is given of the aluminium automotive sheet alloys currently used for outer, inner and structural applications. This includes:

- the typical alloy compositions of the car body sheet alloys
- a comparison of the mechanical (strength and age hardenability) and formability parameters
- some information on corrosion resistance
- general alloy selection criteria for different applications.

Lightweight aluminium sheet is being used for automotive applications due to a number of benefits:

- aluminium offers high potential for weight saving, improving vehicle fuel efficiency and handling characteristics, while reducing at the same time the total greenhouse gas emissions throughout the life of the vehicle
- the metal is easily and widely recycled, saving energy and raw materials
- aluminium will absorb the same amount of crash energy as steel, at just over half the weight.

Different aluminium alloys are used to produce the various automotive sheet panels required to produce a vehicle (Furrer and Bloeck, 2009). The reason is that different applications require a different combination of properties.

85

In general, both AlMgSi (AA6xxx) and AlMg (AA5xxx) alloys are used to produce the body-in-white (BIW) structure as well as the closure panels. In Table 5.1 an overview is given on the advantages of AA6xxx and AA5xxx alloys and the main criteria for alloy selection. These two alloy families exhibit both good strength and formability properties, and excellent corrosion resistance. Furthermore they can be processed in a very similar way to steel through a vehicle manufacturing facility (stamping, joining, assembly, surface treatment), allowing a relatively easy material substitution to be carried out to produce a conventional vehicle structure with approximately 40% mass reduction compared to a steel structure. The applicable forming, joining and surface treatment techniques are the same, although some adjustments of the processing conditions may be necessary to adapt to the different material characteristics. In addition, specific joining methods such as self-pierce riveting and spot friction are more suitable for joining of aluminium automotive sheet (Briskham *et al.*, 2006).

5.2   Sheet alloys for outer applications

For outer body panel applications, AA6xxx alloys (AlMgSi) are almost exclusively used. They show a homogeneous surface quality after forming since they do not develop stretcher strain markings, as would be the case during forming of high Mg containing AA5xxx alloys (see Section 5.3). A specific advantage of AlMgSi alloys is their age-hardening behaviour. The precipitation sequence, which leads to a significant strength increase in a commercially interesting temperature range, has been intensively examined (Edwards *et al.*, 1998). The material is typically supplied in the annealed

Table 5.1 Advantages of AA6xxx and AA5xxx alloys and the main criteria for alloy selection

AA6xxx age hardenable	Criteria for alloy selection	AA5xxx non age hardenable
++	Formability	+++
+++	Hem flanging	+++
++	Corrosion resistance	+++
+++	Heat resistance at $\geq 65°C$	Has to be checked for alloys with nominal >3% Mg
+++	Crash performance	+++
+++	Visual surface appearance of panels	Stretcher strain markings
+++	Strength, i.e. potential for downgauging	+
Increase	Strength change during heat treatments	Decrease
+++	Recycling, i.e. same alloy family for inners and outers	+

+++ = excellent
++ = good
+ = moderate

and solution heat treated condition (T4 temper), which exhibits very good formability. The cold deformation during the forming operation and the heat treatment applied during the subsequent paint bake cycle lead to a significant strength increase. A high material strength in the final panel is of great importance to achieve the required panel dent resistance, and hence to allow down gauging of the material thickness with a corresponding weight reduction.

The selection of the alloy depends on the property requirements. The standard AlMgSi alloy that has been used in Europe for more than 25 years is EN AW-6016 (Anticorodal®-121, Ac-121). For outer panel applications with more stringent requirements, specific AlMgSi alloys have been developed offering enhanced properties. For production of panels that require a very good formability and an extremely good bendability of the sheet, e.g. to allow sharp hemming even in highly pre-deformed areas to ensure small visual gaps between adjacent panels, the alloy EN AW-6014 has been developed, which is marketed as Anticorodal®-170 (Ac-170). For panels where a high strength is the main requirement, alloys such as EN AW-6111 and Ac-140 offer an increased strength in the as-supplied condition and a stronger age hardening response during heat treatment. In Table 5.2 typical chemical compositions are given for the main AlMgSi alloys for outer panel applications. Table 5.3 shows the typical mechanical properties in the as-supplied T4 temper and data on the age hardening by heat treatments. In Fig. 5.1 the bending factor f of the different alloys is shown as a function of pre-straining before bending.

Strengthening mechanisms for AlMgSi alloys

The strength of the AlMgSi alloys in the T4 temper is determined by the relative amounts of the main alloying elements magnesium and silicon, but the content of manganese and copper also has a significant effect. In addition, the applied processing conditions, especially the hot rolling and solution heat treatment parameters, have a substantial influence on the sheet

Table 5.2 Typical chemical composition of the main AlMgSi alloys for outer panel applications

Alloy*	Novelis trade name	Content in wt%				
		Si	Fe	Cu	Mn	Mg
6016	Ac-121	1.1	0.20	0.08	0.07	0.4
6014	Ac-170	0.6	0.20	0.13	0.07	0.6
6016	Ac-140	1.0	0.20	0.04	0.07	0.5
6111	–	0.6	0.20	0.7	0.15	0.7

*Designation according to AA or EN AW-, respectively

Table 5.3 Typical mechanical properties of the main AlMgSi alloys for outer panel applications in the as supplied condition and after straining and heat treatment to simulate forming and electro-coat curing

Alloy*	Novelis trade name	Temper (T4 or PX**)	0.2% PS (MPa)	UTS (MPa)	Ag (%)	A80 (%)	n_m	r_m	0.2% PS after 2% straining + 185°C × 20 min (MPa)
6016	Ac-121	T4	95	205	22	26	0.30	0.70	140
6016	Ac-121	PX	100	210	22	26	0.29	0.84	215
6014	Ac-170	PX	90	195	22	25	0.29	0.68	220
6014	Ac-140	PX	115	230	22	26	0.29	0.70	240
6111	–	PX	125	245	21	25	0.27	0.65	255

*Designation according to AA or EN AW-, respectively
**PX = pre-aged

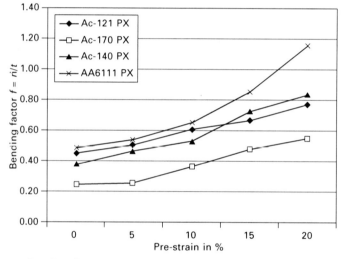

5.1 Bending factor *f* of AlMgSi outer sheet alloys after different degrees of pre-straining. 180° transverse bending test carried out after 3 months RT storage after solution heat treatment.

strength, although their influence on formability and bending performance is even more important. For AlMgSi alloys the strength increase due to work hardening, resulting from forming operations, is somewhat smaller than for AlMg alloys. However, the big advantage of the AlMgSi alloy system is the large age hardening potential during thermal treatments, e.g. during the paint bake cycle. For a solutionised AlMgSi alloy, a combination of cold work (such as may be carried out in the stamping operation) followed by a thermal treatment in the temperature range 165°C to 210°C (such as the paint bake cycle), leads to a strong increase in the in-service panel strength. The paint bake response is caused by precipitation hardening, i.e. the nucleation and growth of fine, closely distributed particles of the precipitating phase. These particles hinder the movement of the dislocations which is necessary for a deformation process. The largest strength increase of the pressed outer panels takes place during the curing of the first lacquer layer, the cathodic electro-coating, which is usually carried out at temperatures between 175 °C and 185 °C. The curing of subsequent lacquer layers is carried out at lower temperatures, and as a result they add only a small amount of additional strength.

In order to exploit the full age hardening potential of AlMgSi alloys during lacquer baking, i.e. without the use of a separate dedicated heat treatment of the body in white, and also noting the current trend to reduce lacquer curing temperatures and/or times, a so-called 'pre-ageing' treatment was developed. Immediately after the solution heat treatment and rapid quenching process, the aluminium strip is subjected to a thermal treatment at elevated temperatures which promotes the precipitation process in a subsequent thermal treatment. As a result, the precipitation kinetic during electro-coat curing is significantly accelerated and precipitation hardening is shifted to lower temperatures. Such a 'pre-aging' step can be applied to all AlMgSi alloys.

Novelis alloys that have been subjected to a pre-ageing treatment have the designation PX. The alloy Ac-170 PX is characterised by exhibiting excellent formability and hemming performance, very good age hardenability, excellent corrosion resistance and good stability of the mechanical properties during room temperature storage. Due to these combined properties, this alloy is now predominantly used for critical outer hang-on panels and can also be recommended for demanding inner applications. Typical areas of application are hood outers, fenders, door outers, decklids and critical inner panels. A comparison of the age hardenability of some AlMgSi alloys in the T4 and pre-aged PX temper is shown in Fig. 5.2.

Corrosion performance of AlMgSi alloys

The primary corrosion mechanism in AlMgSi alloys, when exposed to corrosive conditions, is intergranular corrosion. Intergranular corrosion is the selective dissolution of the grain boundary zone, while the bulk grain is not attacked. Intergranular corrosion is caused by the action of micro-galvanic cells at the grain boundaries. Grain boundaries are preferred sites for segregation and precipitation, which makes them physically and chemically different from the matrix. Furthermore, a zone adjacent to the grain boundary is depleted of the solute elements. Consequently, a 'galvanic cell' is formed. If, for example, the grain boundary precipitates are electro-chemically nobler than the aluminium matrix, the depleted zone becomes electrochemically active, i.e. is preferentially dissolved. But also the opposite case is possible (Svenningsen, 2003). The susceptibility to intergranular corrosion is influenced by:

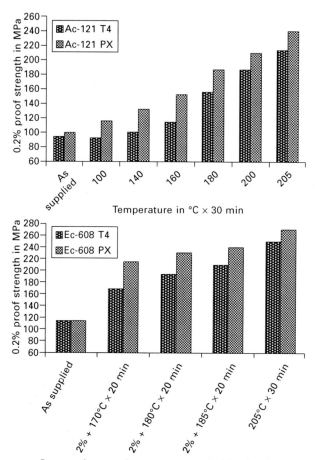

5.2 Age hardenability of the AlMgSi alloys Ac-121 and Ec-608 in the T4 and pre-aged PX temper.

- *The alloy composition*: An increased Cu content makes the material more sensitive to a corrosion attack (Meissner, 1992). This is explained by the precipitation of Al_2Cu particles at the grain boundaries, being more noble than the aluminium matrix.
- *The thermo-mechanical processing*: Material in the solution treated temper (T4) is more corrosion resistant than cold formed and heat treated material.

However, it should be noted that Cu-free 6xxx-series alloys commonly used for automotive sheets resist this type of corrosion. Care has to be taken only for materials with an extreme Si/Mg ratio (> 3) as well as with Cu-containing alloys (Zhan *et al.*, 2008). Copper contents above 0.3% increase the strength of 6xxx series alloys but may also affect the stability against intergranular corrosion. The higher the Cu content, the more attention is needed to avoid this kind of corrosion. Components made out of Cu-containing 6xxx materials should be tested against intergranular corrosion or provided with a proper corrosion protection. This is the case, for example, for BIW applications where a suitable coating system (pre-treatment and lacquer) is used. The alloy EN AW-6111 with nominally 0.7 weight% Cu is extensively used for outer body panels with good long-term experience.

5.3 Sheet alloys for inner closure panels and structural applications

For inner closure panels and structural applications, both AlMg and AlMgSi alloys are used. For some considerable time, only the AlMg alloys AA5754 and AA5182 were most commonly used. Both alloys show very good formability characteristics and enable the fabrication of panels with highly demanding geometries. In a crash situation, these alloys exhibit good energy absorption capability which hinders early failure of the panels. However, AlMgSi alloys are also often selected for visible inner closure panels due to the fact that during the forming process AlMg alloys can develop two types of surface irregularities:

1. Type A stretcher strain markings (yield point elongation (YPE), Portevin le Chatelier effect) appearing up to about 1% elongation as flame like patterns with a maximum roughness of about 75 μm. These are highly visible after painting. It is possible to avoid the occurrence of the type A stretcher strain markings by a light levelling operation of the annealed strip (small cold deformation). This treatment will, however, result in a small decrease of the material formability.

2. Type B stretcher strain markings (serrated flow) appearing as parallel lines at about 55° to the stretching direction with a maximum roughness of approximately 10 μm. During forming, they start to occur above 2–3% elongation, depending upon the Mg content and the grain size. In most cases, these lines will remain visible after painting. In order to achieve a homogeneous surface appearance as required for visible panels, the formed panels would have to be ground which can have a negative effect on the corrosion resistance of the painted material.

AlMg alloys are not age hardenable. Their strength is determined by:

•   *The chemical composition*: The Mg content has the main effect, but also additions of Mn and Cu lead to a strength increase. The effect of the alloying elements Mg, Mn and Cu is due to solid solution strengthening. The substitution of aluminium atoms by solute atoms leads to a distortion of the Al lattice. During sheet forming, the movement of dislocations is hindered (e.g. dislocation pinning by Mg atoms), leading to an increase of the yield strength. Mn additions can also form fine precipitates and increase the material strength by dispersoid strengthening and the resulting grain refining effect. A unique effect in AlMg alloys is that both strength and elongation to fracture are increased with an increase of the Mg content.

- *The sheet grain size*: In AA5xxx alloys, the grain size has a much stronger effect on the strength than in other aluminium alloys. During processing of the sheet material the grain size can be influenced by the level of cold work before recrystallisation and by the final annealing practice. An increase of the degree of cold work leads to a finer grain structure after the final anneal and thus to an increased strength.
- *The degree of cold work*: Any cold deformation (e.g. during levelling and forming of the panels) leads to work hardening.
- *The heat treatment of the formed panels during adhesive and paint curing*: During these treatments, usually a softening of the AlMg materials occurs. The high diffusivity of the Mg atoms allows a relatively easy re-arrangement (recovery) of the dislocations and consequently a softening of the material. The degree of softening increases with increasing pre-strain and increasing time at temperature because the stored elastic energy resulting from high work hardening provides a large driving force for thermal recovery and softening.

5.3.1   Corrosion performance of AlMg alloys

AlMg alloys with a Mg content up to nominal 3 wt% are very resistant against corrosion attack. If AA5xxx alloys with a higher Mg content are exposed for an extended time period to temperatures ranging from 65 °C up to 150–200°C, they may become susceptible to various forms of intergranular corrosion in a corrosive environment. The susceptibility to intergranular attack results from the precipitation of a 'continuous' film of β-Mg_5Al_8 phase at grain boundaries, a process called thermal sensitization (Davenport *et al.*, 2006). The β-phase is fairly slow to precipitate, but during prolonged service at elevated temperatures, β-phase precipitates nucleate and grow on grain boundaries until eventually a near-continuous grain boundary film is formed. The rate of precipitation will be accelerated by prior cold deformation, as the increased dislocation density assists the diffusion of Mg atoms.

The β-phase is highly anodic relative to the aluminium matrix, and renders the alloy susceptible to intergranular corrosion by dissolution of the grain boundary phase. Two forms of intergranular corrosion are observed in AlMg alloys in a corrosive environment, namely pure intergranular corrosion (in fully recrystallised material) and exfoliation corrosion. A heavily deformed sheet that has not undergone complete recrystallisation may exhibit flattened pancake shape grains. If such a grain structure is present, intergranular corrosion may take the form of exfoliation where grains flake off the surface as corrosion progresses. Exfoliation corrosion requires the combination of a pancake shaped grain, sensitisation to produce a continuous film of β-Mg_5Al_8 phase at grain boundaries and a corrosive environment.

Intergranular corrosion attack is most critical since corrosion induced cracking under tensile stresses (external loads or internal residual stresses) can occur. The applied stress may result from loading of the aluminium structure or from residual stresses, for example, in the vicinity of a weld. When the use of AlMg alloys with higher than 3 wt% Mg is desired for structural applications, consultation with the material producer is recommended and their applicability must be evaluated in detail. For safety critical parts, the thermal exposure of the part during its lifetime should be known and preferably a realistic, full component test should be performed.

5.3.2   AlMgSi alloys used for inner and structural applications

The use of AlMgSi alloys for inner and structural applications has the following advantages:

- no occurrence of stretcher strain markings during forming which is especially desired for visible inner panels
- the age hardening effect offers the possibility for down gauging in strength dominated areas, and thus weight savings
- easier recycling of aluminium hang-on parts dismantled from end-of-life vehicles due to the use of material from the same alloy family for outer and inner panels
- no sensitivity to corrosion induced cracking under tensile stresses in heat exposed vehicle areas.

The following alloys have been developed to fulfil the specific requirements for inner and structural applications:

- *Ecodal®-608 (EN AW-6181A)*: This Cu-free AlMgSi alloy for inner applications has been developed to allow the use of recycled aluminium. Ecodal-608 is recommended for uni-alloy solutions, i.e. the use of AlMgSi alloys for both the outer and inner panels with advantages in relation to corrosion performance and recycling. In the T4 temper, Ecodal-608 is relatively soft and shows similar formability to Anticorodal®-121 (EN AW-6016). However, in comparison to Ac-121, Ecodal-608 has a higher strength level after age hardening. Therefore Ecodal-608 offers opportunities for down gauging. However, due to its recycling-friendly composition and limited hem flanging capability, Ecodal-608 is not recommended for outer panel applications, but primarily used for visible inner panels of hang-on parts, e.g. hoods and tailgates.

- *Anticorodal®-118 (Ac-118, EN AW-6501)*: Ac-118 is a relatively soft and highly formable alloy. During heat treatment, e.g. lacquer curing, only a small age hardening effect takes place. Hood inner panels are an important application of this alloy. In combination with a specific inner panel geometry, a closely controlled energy absorption capability is achieved giving improved pedestrian protection. In modern hood design, often a combination of Ac-118 is used for the inner and Ac-170 for the outer panel. Due to its excellent formability, Ac-118 is also of interest for highly deep-drawn components such as door inner panels.

- *Anticorodal®-300 (Ac-300, EN AW-6014)*: Ac-300 exhibits an outstanding performance under crash loads, compare Fig. 5.3. This alloy folds without the formation of cracks, does not tend to fragment during fracture and exhibits a significantly higher mass-specific energy absorption capacity compared to standard steel grades.

5.3 Crash performance of Ac-300, photo of a riveted tube of 2.5 mm Ac-300 heat treated at 160°C × 12 h to achieve T6 temper.

The typical chemical composition and mechanical properties of the alloys described above are given in Tables 5.4 and 5.5. In Figure 5.4 the effect of heat treatment applied during lacquer curing is shown on the 0.2% proof strength of AlMgSi and AlMg alloys.

Table 5.4 Typical chemical composition of AlMgSi alloys for inner and structural applications

Alloy*	Novelis trade name	Content in weight%				
		Si	Fe	Cu	Mn	Mg
6181A	Ec-608	0.9	0.25	0.10	0.12	0.8
6501	Ac-118	0.4	0.20	0.08	0.07	0.4
6014	Ac-300	0.6	0.20	0.13	0.07	0.6

*Designation according to AA or EN AW-, respectively

Table 5.5 Typical mechanical properties of AlMgSi alloys for inner and structural applications in the as supplied condition and after straining and heat treatment to simulate forming and electro-coat curing

Alloy*	Novelis trade name	Temper	0.2% PS (MPa)	UTS (MPa)	Ag (%)	A80 (%)	n_m	r_m	0.2% PS after 2% straining + 185°C × 20 min (MPa)
6181A	Ec-608	T4	105	215	22	24	0.30	0.63	210
6501	Ac-118	T4	70	135	22	26	0.40	0.70	100
6014	Ac-300	T4	110	195	19	22	0.25	0.65	180

* Designation according to AA or EN AW-, respectively

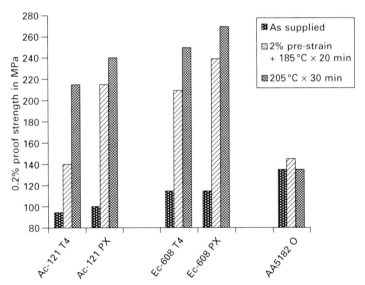

5.4 Effect of heat treatment during lacquer curing on the 0.2% PS of the AlMgSi alloys Ac-121 and Ec-608 and the AlMg alloy AA5182.

5.4   Fusion alloys

Novelis Fusion™ is a new casting technology that produces simultaneously a high quality rolling ingot with a core of one aluminium alloy, combined with surface layers of a different aluminium alloy composition. It offers a cost-efficient way to independently engineer the surface and core material characteristics of an aluminium sheet while maintaining a consistent high quality interface between both layers. The clad thickness of each surface layer is typically 10% of the total sheet gauge. Thus, the Novelis Fusion technology allows the production of previously difficult or impossible combinations of aluminium alloys for strength, formability, and surface finish, and enables the supply of superior products both for existing and new markets.

The new technology is based on the conventional direct chill (DC) ingot casting method, but includes a secondary heat-removing chamber. Multiple coolant flows and liquid metal streams are controlled by a series of flow control and metal level sensors which guarantee the necessary thermal and mechanical boundary conditions during solidification. Simultaneous casting of the different alloy layers into a single ingot leads to the formation of a truly metallurgical bond between the layers. The planar interface is essentially free from oxides, porosity or other defects and there is little indication of penetration of the core alloy into the cladding. The clad ingot is then processed using the traditional hot and cold rolling steps, similar to the production of monolithic sheet alloys.

A range of Fusion materials have been developed for automotive applications:

- *Fusion™ AF350*: This highly formable multi-alloy sheet was developed for applications with very high deep drawing requirements, e.g. door inner panels, and may be also used for any component having similar requirements. AF350 is a combination of an AlMg5 core material with high formability, and an AlMg1 (EN AW-5005) clad surface layer that protects the material against intergranular corrosion attack, and also enhances the adhesive bonding performance of the sheet. The high Mg content of the core, required for high formability and strength, makes this clad material susceptible to the generation of stretcher strain marks during press forming, which will remain visible even after lacquering. Therefore, application is recommended only for inner parts. Due to its excellent formability, AF350 offers the possibility for new design solutions that have not been achievable to date with standard monolithic alloy sheets:
  - unique opportunity to design and stamp highly complicated parts, e.g. a one-piece door inner with a window frame
  - reduced manufacturing cost due to lower assembly costs compared to a multi-part design and consequently also an additional weight reduction potential.

- *Fusion^{TM} AS250*: The development target was structural panels that require high strength and superior crash energy absorption capacity. AS250 exhibits good formability characteristics and may also be used for outer applications. It consists of a high strength core material (corresponding to EN AW-6111 with a reduced Cu content) which is clad with Ac-170, an alloy specifically developed for tight hemming. In addition, as a result of the clad layer composition, the material exhibits a very good corrosion resistance. Due to the increased material strength in the age hardened condition, AS250 offers a sheet thickness reduction potential compared to a monolithic aluminium alloy solution. Figure 5.5 shows the typical microstructure of the core and clad of AS250 at final gauge.
- *Fusion^{TM} AS300*: This alloy was developed to enable a further strength increase compared to AS250, but to maintain the excellent hemming performance and very good corrosion resistance. AS300 consists of a high strength core alloy (EN AW-6111 with increased Cu content) which is clad with Ac-170. The increased strength of AS300 increases the resistance to local denting and hence the potential to down gauge panels resulting in a further weight reduction on automotive outer panels and reinforcement parts.

The typical mechanical properties of the mentioned Fusion materials are shown in Table 5.6.

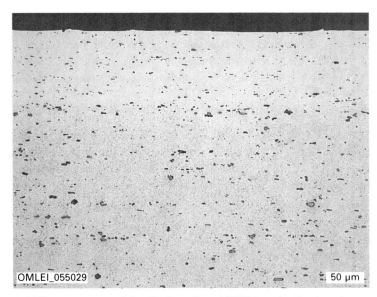

OMLEI_055029   50 µm

5.5 Microstructure of the Fusion alloy AS250 longitudinal to the rolling direction showing the clad and core material and the interface. Etching of the metallographic cross section was done in NaOH.

Table 5.6 Typical mechanical properties of Fusion alloys for automotive applications in the as supplied condition and after straining and heat treatment to simulate forming and electro-coat curing

Novelis trade name	Temper	0.2% PS (MPa)	UTS (MPa)	Ag (%)	A80 (%)	n_m	r_m	0.2% PS after 2% straining + 185°C × 20 min (MPa)
Fusion™ AF350	O	130	275	25	28	0.33	0.73	150
Fusion™ AS250	PX	135	250	21	24	0.26	0.56	265

5.5   Surface treatment of the aluminium strip

5.5.1   Surface texturing

Currently used surface topographies of automotive aluminium sheets, generated during the last cold rolling pass, are described below together with their associated advantages.

In Europe almost all car body sheets are provided with a special surface topography. The advantages of textured sheet surfaces compared to the standard mill finish surface are:

- a more equal distribution and better adhesion of the lubricants applied to the strip surface
- improved performance of the sheet or strip material during transport, handling and storage, i.e. reduced risk of the formation of surface defects due to fretting
- easier automatic de-stacking of the sheets in the press shop
- improved formability
- homogeneous class A surface appearance after painting.

The desired surface topography is transferred onto the strip surface, using work rolls with a specifically prepared surface structure, during the last cold rolling pass. The EDT (electro-discharge texturing) structure, where the work roll surface is textured by means of spark erosion, is the most frequently used surface topography in Europe. EDT textured rolls produce a strip surface with relatively flat hollows which act as lubricant pockets during the sheet forming process and thus improve the formability of the sheet. Figure 5.6 shows the roughness value (R_a) and the closed void volume (V_{cl}) for mill finish and EDT topography. SEM micrographs of these two topographies are given in Fig. 5.7. The stochastic, fine surface structure prevents an impairment of the visual appearance of the formed panel after painting. Other sheet surface texturing methods include the roughening of the roll surface by shot blasting (used in Japan for aluminium automotive sheet), laser or electron beam texturing of the roll surface, or the application of a structured chromium plating to the roll surface, called Topocrom®. However, these latter techniques are not used for aluminium automotive sheet in Europe.

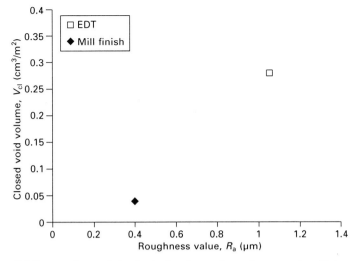

5.6 Comparison of the typical values of roughness value (R_a) and closed void volume (V_{cl}) of mill finish and EDT topography on automotive sheet.

5.5.2   Chemical strip treatments

The reasons why strip surface treatments are carried out are discussed in this section. The different types of surface treatments, i.e. degreasing/etching, conversion treatments (based on either Ti- or TiZr-fluoride), chemical treatment with silicate containing agent, and thin anodised films are described, together with the advantages of a strip surface treatment.

The surface of aluminium is, due to the immediate reaction with oxygen in the air, always covered with a thin layer of aluminium oxide. The natural oxide film follows exactly the material surface topography, it has a thickness of a few nanometres, it is transparent, amorphous, tight (prevents the further access of oxygen to the Al metal) and stable in the pH range of pH 4 to 8.5. The presence of this oxide film is the reason for the good corrosion resistance of aluminium. However, one disadvantage of the natural oxide film is that it is not perfect. It can show some thickness variations and exhibit defects such as pores and fine cracks, the oxide can be disturbed in areas where intermetallic particles are present in the adjacent aluminium metal, and it can contain surface contaminations, e.g. residues of rolling oils. Furthermore, solute Mg atoms can diffuse from the matrix into the surface oxide layer, leading to the formation and enrichment of magnesium oxide at the sheet surface. The diffusion of the Mg atoms to the surface is relatively fast and occurs already during heat treatments at temperatures >200 °C since the Mg atoms move preferentially along the grain boundaries. Similar effects can be observed for other elements such as Li, Na, Be, and Ca, although these impurity

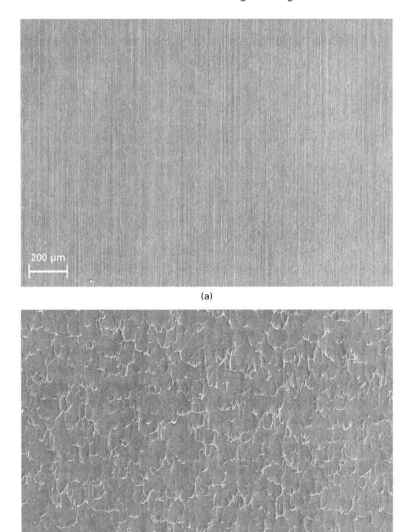

5.7 (a) Mill finish and (b) EDT topography on automotive sheet: SEM micrographs.

elements are closely controlled and limited to very low concentrations. This affects the characteristics of the oxide layer, i.e. makes it more hygroscopic, less corrosion resistant and affects the adhesion to adhesives (reduces the long-term performance of adhesive bonds), primers and lacquers. For these reasons, the aluminium strip surface is often subjected to chemical or electrochemical treatments in order to remove the inhomogeneous natural Al oxide layer, and to generate a new, homogeneous oxide film that is free of Mg oxide enrichment (Textor *et al.*, 1995).

Alkaline and acidic cleaning/etching

Prior to the application of special surface treatments onto the strip surface, a cleaning/etching step is usually carried out to remove residues of rolling oil, aluminium debris generated during rolling and the natural inhomogeneous oxide layer. After an alkaline cleaning step, an acidic cleaning step must be always carried out to remove any smut layer. For the acidic cleaning step, sulphuric acid or a mixture of acids, e.g. sulphuric and hydrofluoric acid or sulphuric and phosphoric acid is often used. Also, there are strip treatment lines which use electrolytic cleaning in a phosphoric acid electrolyte.

Conversion treatments

Conversion treatments are carried out using agents based on either titanium fluoride or a mixture of titanium and zirconium fluoride. During this kind of chemical treatment, a modified mixed oxide film containing titanium and zirconium ions is formed which offers a good adhesion to organic compounds, and thus leads to an improved long term stability of adhesive bonds (Bloeck *et al.*, 1999). The modified oxide layer is homogeneous, stable and very thin, i.e. in the nanometre range. Sheet with this kind of surface treatment shows a good and homogeneous performance during welding. Also the very thin conversion layers have no negative effects on the formability of the material or its behaviour during Zn-phosphating.

Thin anodised films

Thin anodised films (TAFs) are generated using an AC or DC powered electrolytic process and have several advantages over conventional chemical cleaning and pre-treatment methods. Firstly, TAFs are composed entirely of aluminium oxide and so provide an environmentally attractive alternative to chromium or other transition metal based pre-treatments. In addition, the thickness and morphology of TAFs can be directly and accurately controlled by varying key parameters such as cell voltage and applied current. The formed oxide film consists of an amorphous barrier layer improving the corrosion resistance of the material, and a porous filament layer giving an excellent adhesion to adhesives, primers and lacquers. Thin anodised films can be 'tailor made' to give excellent long-term stability of adhesive bonds. For structural bonding, a film thickness in the range of 80 nm to 120 nm is recommended with a barrier layer of 30 nm to 40 nm and a filament layer of 50 nm to 80 nm. Figure 5.8 shows a TEM micrograph of a microtome cut through a TAF layer on automotive sheet. Thin anodised films in the quoted thickness range have no negative effects on the material performance in the car manufacturing plant, i.e. they do not affect sheet formability, joining characteristics, surface appearance or corrosion resistance after painting. During a typical layer forming Zn-phosphating treatment of the BIW TAF layers of the mentioned thickness are removed and a homogeneous Zn-phosphate layer is formed. TAF treated sheet can be used for inner and outer sheet applications.

Pre-treatment 2 (PT2)

PT2 was developed by Novelis as a pre-treatment for structural adhesive bonding of aluminium sheet and is part of the Aluminium Vehicle Technology System (see Section 5.5.4). PT2 is a no rinse, chrome and fluoride-free surface treatment for aluminium strip. It is non toxic and is applied on the strip surface as an aqueous suspension of colloidal silicate, with some additions required for film formation and wettability. The generated surface film follows exactly the strip surface topography and has a thickness of about 50 nm to

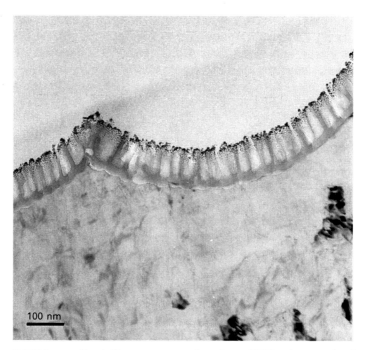

5.8 Thin anodised film on automotive sheet: TEM micrograph of a microtome cut.

100 nm (see Fig. 5.9). Apart from providing excellent long-term stability of adhesive bonds, PT2 offers the following advantages:

- PT2 coated sheet can be welded using spot welding, MIG or laser welding
- PT2 shows high long-term stability of the coated strip, up to 12 months storage time.

Due to the relatively high thickness of the PT2 film, the Zn-phosphating process, carried out prior to the electro-coating, has to be modified. PT2 coated sheet is used for inner and structural sheet applications, but is not recommended for outer panel applications with very high requirements on final painted surface appearance. This is due to the PT2 layer thickness which may affect the homogeneity of the surface appearance after Zn-phosphating and lacquering. PT2 is a pre-treatment developed for coating of strip and not for application on extrusions or castings.

In Figs 5.10 and 5.11 the tensile shear strength of adhesive bonds produced using aluminium sheet is shown plotted as a function of different kinds of surface treatments when subjected to corrosion exposure.

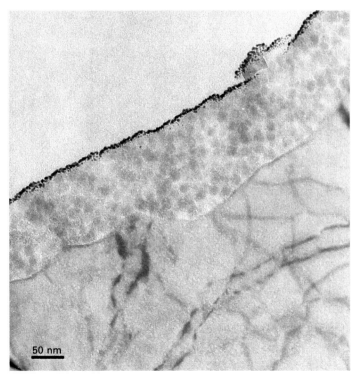

5.9 PT2 coating on automotive sheet: TEM micrograph of a microtome cut.

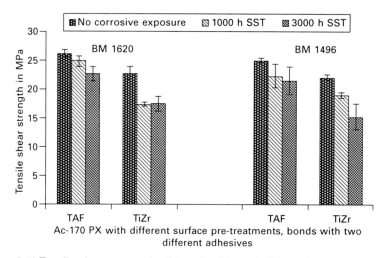

5.10 Tensile shear strength of bonds of Ac-170 PX subjected to the salt spray test according to DIN 50021 (35°C) for 1000 and 3000 hours. Comparison of TAF and TiZr-based pre-treatment and bonds with two different 1K epoxy adhesives.

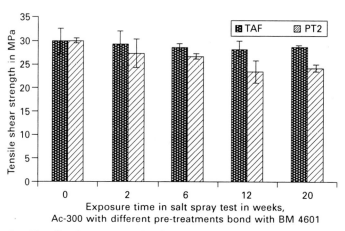

5.11 Tensile shear strength of bonds of Ac-300, 2.5 mm, subjected to the salt spray test according to DIN 50021 (but at 43°C) up to 20 weeks. Pre-treatments: TAF, PT2. Adhesive: BM 4601.

5.5.3   Lubrication

The advantages of using lubricated strip are described below.

Different kinds of lubrication can be applied to the strip surface depending upon customer requirements:

- A thin film of a suitable lubricant can be applied to protect the surface of the strip or the cut panels against corrosion and friction effects during transport, referred to as fretting. Fretting can lead to surface damage. Before forming in the press shop, a proper press forming lubricant has to be applied to the blank.

- A film of a lubricating oil that can be used also for forming, eliminating the need to apply additional lubrication at the press shop. This option is suitable when the material is delivered in coil form. However, it must be noted that oil can redistribute in the coiled state resulting in an inhomogeneous distribution. When cut panels are supplied, lubrication with oil needs to be considered carefully due to potential issues related to de-stacking of the panels in the press shop.

- A film of a dry lubricant suitable for press forming with a typical coating weight in the range of 0.8 g/m^2 to 1.2 g/m^2 depending on the complexity of the panel to be formed. The trend is to use mineral oil based dry lubricants that are applicable for both aluminium and steel sheet. The lubricants are applied at temperatures of about 60 °C to 70 °C, and solidify at about 50°C. The dry lubricated sheet material shows good formability without additional lubrication in the press shop. In general, the dry lubricants are not removed from the stamped panels. Consequently, further requirements that the lubricants have to fulfil are:

   - Compatibility with any adhesives applied to the formed panels during assembly in order to create safe bonded joints, without the need for a degreasing step.

   - Electro-coat compatibility. In a 'lean production' process the lubricant is removed from the assembled BIW only in the alkaline degreasing step, before Zn-phosphating. This means that the lubricant has to be easily removable, even after deformation and heat treatment (e.g. pre-curing of the adhesive). It is known that these process steps can affect the removability of lubricants. Even if the panels are thoroughly degreased in the alkaline degreasing step, traces of lubricant present in crevices and on the inner part of the BIW can still be transferred into the electro-coat bath. If the lubricant is not fully compatible with the electro-coat, craters or pimples can develop on the panel surface during e-coat curing.

Up to now, no dry lubricants are available which allow MIG, TIG and/or laser welding of the material in the lubricated condition (at a lubricant film thickness required for forming) in a good weld quality without formation of porosity, without a pre-cleaning step.

5.5.4   Aluminium Vehicle Technology (AVT) system for structural bonding of aluminium sheet

The key features of the AVT system (combination of alloy + strip cleaning + chemical surface treatment + lubrication + adhesive selection) are described below.

The AVT system has been developed for the production of full aluminium car bodies. The different elements are:

- *Alloys*: traditionally AA5754 and also AA6111 have been used for the structural parts of the car body. Nowadays other alloys such as Ac-300 and Fusion materials are also used.
- *Chemical surface treatment*: the material is acid cleaned, preferably electrolytically, followed by the application of PT2
- *Lubrication of the strip surface*: a hot melt, ALO70, is applied with a coating weight ranging from 1.5 g/m^2 to 2 g/m^2 depending on the panel complexity. This lubricant offers very good formability of the sheet and it is also suitable for resistance spot welding.

The supplied aluminium panels are formed without additional lubrication. The panels are then joined (without a prior cleaning step) by a combination of adhesive bonding and riveting. The adhesive used is a 1-K epoxy adhesive, Betamate 4601 (previously named XD4601), which is compatible with the ALO70 lubricant. As an alternative, a combination of adhesive bonding and resistance spot welding can be used. The combination of adhesive bonding together with either riveting or spot welding significantly increases the fatigue life endurance, and enables a significant weight reduction to be achieved whilst maintaining strength and increasing stiffness of the body structure.

5.5.5   Primer coating of the strip

Both conductive and non conductive primers are available for strip coating, offering the potential for a replacement of the electro-coat or the electro-coat and filler. The advantages of these types of organic coatings are described and the limitations discussed below.

Modern strip surface treatment lines offer the possibility to apply an organic coating, subsequent to a suitable chemical pre-treatment step which is required to ensure adequate primer adhesion and corrosion resistance. Conductive primers are of particular benefit for protection against galvanic corrosion in mixed metal constructions, i.e. aluminium and steel and/or galvanised steel. Other advantages of material coated with a conductive primer are:

- surface protection during transport and handling
- improved formability, use of less lubricant
- good adhesive bonding characteristics
- reduction of the aluminium surface area that has to be pre-treated in the paint line, e.g. by means of Zn-phosphating, before electro-coating
- good surface preparation for the lacquering steps in the automotive paint line.

The use of non conductive primers offers the possibility to save process steps at the car producer. There are highly formable primers available that allow the replacement of the electro-coat and the filler. This may be of some advantage for certain panels, as e.g. roof panels which are mounted onto the BIW after the electro-coating process step.

Limitations of primer coated sheet that have to be considered are:

- grinding of the formed panels must be avoided
- appropriate joining methods need to be selected, i.e. adhesive bonding and mechanical joining. Welding of primer coated aluminium sheet is not possible.

5.6   Future trends

According to Bassi (2010) further developments concentrate on continuous improvement of the sheet material performance regarding

- formability
- strength and hemming performance in combination
- strength and performance during a crash situation in combination
- achievement of a roping free material quality also after high degrees of straining, which is especially of importance for panels with high complexity
- increased usage of recycled materials, which is of interest regarding material cost reduction.

An increase of the material strength also allows downgauging and thus saving of weight and costs.

The properties of a material can be influenced both by the chemical composition and the processing history. The processing conditions have an effect, e.g. on the mechanical properties, the formability, the hemming performance, the sensitivity to develop roping during forming, the stability of the material properties during room temperature storage and on the surface characteristics. In order to guarantee that the produced materials exhibit properties inside a defined specification, tight process windows have to be assured in the whole production chain ranging from casting, homogenisation of the ingots, hot and cold rolling, annealing and surface treatment steps. Material properties in a narrow range are essential for the process stability during car production and are a pre-requisite for a high volume production at the car manufacturers. In order to accelerate the growth of aluminium sheet usage in automotive applications on a global basis, it is important to increase productivity and cost efficiency. One possibility is to offer standardised alloys instead of materials optimised for each single application. This may lead in the future to a reduction of the number of alloys, to the usage of the same alloy compositions for inner and outer applications and to the usage of the same alloy for vehicle applications with different requirements. In order to achieve these targets the interactive co-operation between the aluminium suppliers and the automotive companies will gain further importance.

5.7   References

Bassi C (2010), 'Automotive lightweight trends and development strategy for aluminium sheet', Automotive Circle International, Car body lightweight design based on aluminium – towards harmonised aluminium alloy usage, 22 September 2010, Jaguar Land Rover Cars Castle Bromwich, Great Britain.

Bloeck M, Millet P and Bassi C (1999), 'In-line surface pretreated aluminium sheet: a key success factor for high volume production of car bodies', VDEh, METEC Congress 99, *International Conference on New Developments in Metallurgical Process Technology*, Düsseldorf, 13–15 June, 423–428.

Briskham P, Blundell N, Han L, Hewitt R, Young K and Boomer D (2006), 'Comparison of self-pierce riveting, resistance spot welding and spot friction joining for aluminium automotive sheet', *SAE Technical Paper*, 2006-01-0774.

Davenport A J, Yuan Y, Ambat R, Connolly B J, Strangwood M, Afseth A and Scamans G M (2006), 'Intergranular corrosion and stress corrosion cracking of sensitised AA5182', *Materials Science Forum*, Aluminium Alloys 2006 – ICAA10, 641–646.

Edwards G A, Stiller K, Dunlop G L and Cooper M J (1998), 'The precipitation sequence in Al-Mg-Si alloys', *Acta Materialia*, 46, 3893–3904.

Furrer P and Bloeck M (2009), *Aluminium Car Body Sheets*, Munich, Süddeutscher Verlag.

Meissner H (1992), 'Einfluss von Kupfer auf das Korrosionsverhalten von AlMgSi-Karosserieblechen', *Aluminium*, 68, 12, 1077–1080.

Svenningsen G (2003), 'Corrosion of aluminium alloys', Poster at the NorLight Conference, Trondheim, 27–28 January.

Textor M, Timm J and Néma P (1995), 'Car body alloys and methods of corrosion protection: Aluminium sheet', *Materialwissenschaft und Werkstofftechnik*, 26, 318–326.

Zhan H, Mol J M C, Hannour F, Zhuang L, Terryn H and de Wit J H W (2008), 'The influence of copper content on intergranular corrosion of model AlMgSi(Cu) alloys', *Materials and Corrosion*, 59, 670–675.

6
High-pressure die-cast (HPDC) aluminium alloys for automotive applications

F. CASAROTTO, A. J. FRANKE and R. FRANKE,
Rheinfelden Alloys GmbH & Co. KG, Germany

Abstract: This chapter aims to trace a short history of the logical steps which led to the development of new aluminium materials for automotive applications. The main focus is on high-pressure die-cast (HPDC) aluminium alloys and the driving factors leading to the current use of premium casting alloys. The chemical composition, mechanical properties and specific features of Silafont®-36, Magsimal®-59 and Castasil®-37 are described, with guidelines provided for their processing and casting. Data from practical experiences are reported, answering the most common questions on primary low-iron ductile alloys. Finally, examples of the applications of these materials are briefly described, underlining the innovative aspect of each component.

Key words: aluminium alloys for automotive applications, innovative light metals, high-pressure die-cast (HPDC) aluminium alloys, low-iron primary alloys, ductile casting alloys.

6.1 Introduction

This chapter provides a short history of the steps leading to to the development of new aluminium materials for automotive applications, with specific emphasis on high-pressure die-casting (HPDCing) and the driving factors leading to the current use of premium casting alloys.

The process of HPDCing has been carried out for more than 100 years. It was originally used just for zinc; today it is the default choice for light metals, when the volume justifies mass production. Engine cradles, suspension and engine parts, cross members and nodes for space-frame constructions are typical examples of products developed through HPDCing. Until the end of the 1990s, developments in HPDCing focussed principally on the equipment. HPDCing was considered to be appropriate for simple casting of low cost secondary alloys and was not treated as a complex high tech process. It was generally assumed that HPDCing could be used only for non-challenging castings of AlSi9Cu3Fe, where in most cases metal handling and ductility were not even considered, and only productivity counted. The soundness of the castings was evaluated using soft criteria, as the final application had no involvement with safety concerns or critical parts. New equipment

109

and presses allowed huge improvements in the casting process, as did an increased understanding of metallurgical process, e.g. in the areas of high-vacuum HPDCing, squeeze casting and semi-solid casting.

From the 1990s onwards, thanks to the boom in the automotive industry, attention progressively shifted to other areas, previously considered 'appendixes' of the core manufacturing process. The need to develop more complex castings was determined by the continuous marketing of new improved car models; this in turn led the foundry engineers to aspire to a higher quality HPDCing process. High-pressure die-castings were suddenly required to be heat-treated, welded, ductile and stronger. Cast components were designed with larger size but with lower wall thickness, in order to save weight. However, light weight design was still subject to strict quality requirements, defined by the constructor's safety criteria. Heat-treatable castings can achieve a wide range of mechanical properties, but presuppose an oxide-, gas- and hydrogen-free structure, in order to prevent the formation of blisters during treatment. Die-lubricant residues must be minimized, as they favour gas pores, especially in a weld-seam during welding. An improved die-design, including the gating system, new die-lubricants, vacuum and venting systems, was thus developed.

During the manufacturing process, this was translated into new procedures, such as the control of the metal, the introduction of modifications and grain refinement. Monitoring of the melt quality and tuning of the post-manufacturing stages has also recently become an important issue. The main R&D output has focused on improving casting alloys through a material science approach. New low-iron ductile HPDC alloys, described in the following paragraphs, were introduced to the market in the 1990s, allowing a revolution in the field, and substituting with aluminium and steel applications.

This phase is still a work-in-progress, as nowadays increasingly high mechanical properties in terms of yield strength, elongation and fatigue strength are bringing these metals close to their limits. Deformability and ductility are essential, as the castings have to absorb the highest amount of energy in a crash.

HPDCing requires high investments in terms of hardware and tooling. Despite increasing quality controls, the 'cost' factor is growing in importance. High production rates, long tool-life, short downtimes and cycle-times are essential for foundry managers today.

The economical production of these high-pressure die-castings calls for the reduction of heat-treatments to a necessary minimum. If possible, the part should work in its as-cast state. This reduces energy consumption and eliminates expensive straightening operations. Additionally, design engineers integrate many different tasks into large multifunctional castings, in order to reduce the number of components and thereby also the assembly costs.

Quality controls and their cost impact, technical and economical feasibility, the choice of the binomial process-material versus performance of the final part are all two sides of the same coin. In most cases the best compromise must be chosen, taking into account the degree of difficulty of the casting.

The main applications for automotive aluminium HPDCings can be classified into four categories on the basis of their function, as follows:

1. engine and auxiliary systems
2. powertrain
3. interiors and electronics
4. chassis and body structure.

Engine parts and their auxiliaries are exposed to low mechanical stresses, almost just static loads; the most important requirements are elevated temperature properties, pressure tightness and wear resistance. There is therefore no need for expensive alternatives to cylinder head blocks, oil sumps, intake manifolds, water pumps and covers, which can be cast in HPDCing with common AlSiCu(Fe)-alloys. Research trends are currently focussing on promising alloy systems, like AlSi7Cu3(Mn, V, Zr, Ti), AlMg3Si(Cu), AlMg3Si(Sc,Zr) and AlCu5Mn(Ni,Ce) [1] [2]. In any case, these are still under evaluation and are not relevant to HPDCing.

Powertrain parts are also mainly subject to moderate non-dynamic stresses, which favours the use of traditional Al-alloys. Magnesium HPDC alloys are a technically valid alternative, in order to save even more weight in transmission cases and boxes, transfer cases, flanges and couplings. For these applications, the choice between aluminium and magnesium is often made on the basis of the current cost of raw materials and therefore of the final component.

Metal parts for use in interiors are mostly cast in magnesium for weight reasons. Instrument panels and dashboard supports, seat-frames and consoles as well as steering wheels and electronic housings are established magnesium HPDCings. The aluminium 'variant' can be considered if special properties are necessary, such as electric conductivity or optical requirements.

The most relevant areas of progress in terms of casting alloys are in the field of structural parts, with particular reference to the chassis and its auxiliaries. Body structure applications are commonly the most demanding in terms of mechanical properties. Highest strengths coupled with highest elongations are required, as the whole construction needs to absorb the greatest amount of energy in case of crash.

Generally speaking, the complexity of a part is driven by its weight, size, geometry and deformation capabilities. While large parts are difficult to manufacture as heavy machinery is needed, the geometry of the castings and its design require advanced know-how and engineering capabilities. Producing ductile die-castings requires additionally foundry skills as well as alloy competences. Table 6.1 summarizes the main applications of light

Table 6.1 General classification of the most common aluminium automotive parts

Part	Complexity		
	Low	Medium	High
Chassis and body structure	**Hand brake modules** Rear suspensions arms Engine mountings Shock absorbers **Transverse control arms** Wipers	Bed plates **Steering columns** Steering knuckles **Trailing links** **Dampers** Light trims	Suspension struts **Steering gear case** Door panels **Front-end**
Powertrain	**Transmission parts** **Gear box covers**	**Transfer case** **Couplings** **Gear flange** **Fastener for differential**	**Bearing boxes** Clutch housings **4-wheel drive** Transmission case
Engine and auxiliary systems	**Cylinder block covers** **General fastener**	**Water pumps** **Suction pipes** **Covers for camshaft** Engine oil sump	Cylinder block Pistons Intake manifold Cylinder head cover
Interiors and electronics	**Seats** **Brackets and mini-consoles**	**Internal fittings** **Steering wheels**	**Instrument panel** Trims **Dashboards** Radio parts

Application also in HPDC Mg alloys shown in bold.

metals in the automotive industry, classifying them on the basis of their difficulty and function.

From a technical point of view the field of application for standard aluminium alloys is quite limited. Parts with a simple requirement profile, often expressed in terms of low weight and a certain yield strength, are commonly cast in AlSi9Cu3(Fe) or similar. An iron content above 0.6–0.8% is used to avoid the sticking of castings onto the die. Composition tolerances are generally kept broad in order to simplify handling in the foundry. Corrosion resistance plays no role in this respect. When a certain ductility, combined with a normal yield strength is required, Mg-free AlSi-alloys such as AlSi12(Fe) or AlSi9 are normally selected. When a compromise is required, the AlSi10Mg(Fe) can be considered the default solution.

Primary low-iron AlSiMg-alloys are the most common choice, with a heat-treatment normally carried out to achieve the final properties required. A full-age hardening (T6) is performed, in order to obtain the maximum strength. Engine cradles or suspension parts are often designed with the benefits of T6 taken into account. Full heat-treatments, such as T4 or T7, are instead conducted if the key properties are elongation and ductility. Space-frame nodes or structural parts are typical applicative examples.

The first alloy used in this field was Silafont®-36, AlSi9MgMn, developed by the German company Aluminium Rheinfelden GmbH and applied in the space-frame nodes of the Audi A8 in 1994 [16]. The alloy was accepted and registered by the American Aluminium Association in 1996, under the designation 365.0. Further details relating to Silafont®-36 will be provided in Section 6.2.

Alternative aluminium materials for body parts are AlMgSi-alloys, to be used in the as-cast state (temper F) or after a single step heat treatment (temper T5 or O). The most common alloy of this family has the nominal composition AlMg5Si2Mn and is commercially known by the brand name Magsimal®-59. The alloy was developed in 1995 by the same company, Aluminium Rheinfelden GmbH. Further information on Magsimal®-59 will be provided in Section 6.3.

Chassis parts normally require yield strengths above 120 MPa and elongation above 10%. These requirements can be fulfilled in most cases by means of AlSiMg-alloys in the T5 state. Bed plates, space-frame nodes and door panels are some standard examples.

One new alloy should also be mentioned, which is tailored to this kind of application and is capable of combining high elongations with moderate yield strength in the as-cast state. Its chemistry corresponds to AlSi9MnMoZr, known by the commercial name Castasil®-37. This is another alloy developed by Aluminium Rheinfelden GmbH, first produced during the late 1990s and put on the market in 2003. Further details of Castasil®-37 will be provided in Section 6.4.

Connecting the aluminium HPDC alloys to the requirement profiles for automotive castings gives the diagram shown in Fig. 6.1. Yield strength and

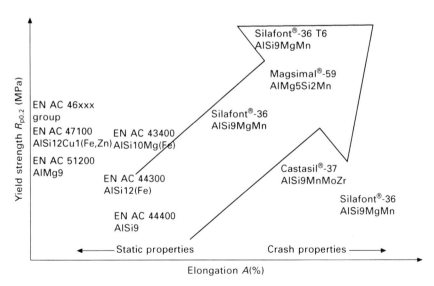

6.1 Ranking aluminium alloys, focussing on the mechanical properties.

elongation have been chosen as the key mechanical properties to qualitatively rank the materials. However, the transition between static properties and crash properties can be defined through an elongation of 4–6%. R&D activities and a scientific material design approach allowed this hurdle to be overcome, which was inconceivable 30 years ago.

Focussing on the new developments in the field of aluminium HPDC alloys requires a closer examination of the high strength and high ductility alloys (see Sections 6.2 and 6.3) and of metals capable of reducing or even eliminating heat-treatment (see Sections 6.3 and 6.4). These new alloys have made possible some applications which would not even have been considered in the past.

The approach adopted in the following sections consists of firstly presenting an overview of these alloys, describing their chemical composition, mechanical properties and particular features. Guidelines for processing and casting are given, as is information useful for the foundry. Some data from practical experiences are reported in order to provide answers to the most common questions about primary low-iron ductile alloys. General notes on after-casting operations are also given, to complete the overview. Finally, examples of the application of these alloys are briefly described, underlining the innovative aspect of each component.

The aim of the authors is not to provide an exhaustive scientific tract of recently developed alloys, but to provide a context for the relevant alloys in the multifaceted world of a modern HPDCing process with some complementary notes on economical aspect.

As a background note to the whole chapter, it should be taken into account that HPDC technology should always be calibrated to a precise target quality level, as shown in Fig. 6.2. Increasing requirements call for a more and more detailed control of the process; fulfilling these requirements means implementing additional measures over eight schematic levels, ranging from a simple dimensionally controlled part up to heat-treatable HPDCings. The main factors with potential for improvement have been found to be air, melt treatments and casting and mould release agents. Suitable HPDC alloys are indicated on the bottom line of Fig. 6.2. Premium casting alloys can be value-added; their potential becomes clear when they are combined with the finest casting process.

6.2   AlSi heat-treatable alloys – Silafont®-36

Before the development of low-iron alloys, parts produced through HPDCing were limited to applications which did not require ductility. As the requirements called for crash capable parts, a new corrosion-resistant alloy for automotive applications was clearly necessary.

In 1994 Silafont®-36 was presented as the first ductile aluminium alloy for

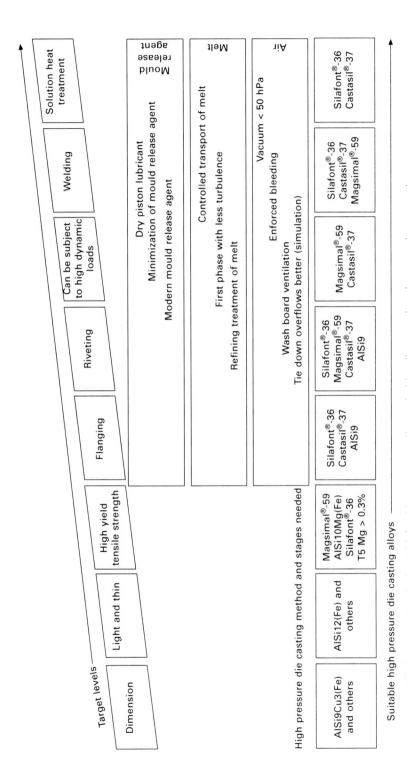

Target levels

| Dimension | Light and thin | High yield tensile strength | Flanging | Riveting | Can be subject to high dynamic loads | Welding | Solution heat treatment |

High pressure die casting method and stages needed

| AlSi9Cu3(Fe) and others | AlSi12(Fe) and others | Magsimal®-59 AlSi10Mg(Fe) Silafont®-36 T5 Mg > 0.3% | Silafont®-36 Castasil®-37 AlSi9 | Silafont®-36 Magsimal®-59 Castasil®-37 AlSi9 | Magsimal®-59 Castasil®-37 | Silafont®-36 Castasil®-37 Magsimal®-59 | Silafont®-36 Castasil®-37 |

Suitable high pressure die casting alloys

Dry piston lubricant
Minimization of mould release agent
Modern mould release agent

Mould release agent

Controlled transport of melt
First phase with less turbulence
Refining treatment of melt

Melt

Vacuum < 50 hPa
Enforced bleeding
Wash board ventilation
Tie down overflows better (simulation)

Air

6.2 Eight target levels for HPDC with corresponding suitable alloys and main control process levers.

© Woodhead Publishing Limited, 2012

HPDCing. This alloy is chemically based on the aluminium-silicon system, which has proven to be the best in terms of castability in HPDCing and has the widest range of applications. The chemical composition of Silafont®-36 is given in Table 6.2.

Silafont® is based on 99.8% pure primary aluminium. It has an average silicon content of 10.5%, providing excellent castability. The iron content is limited to a maximum of 0.15%, in order to avoid the formation of intermetallic β-AlFeSi phases, which are known to be the most detrimental to the dynamic properties and to the elongation depending on their quantity and shape [3]. However, as iron helps in the ejection of the castings from the die, manganese is alloyed, acting as a functional substitute. The best Mn-content has been found to be between 0.5–0.8%; below this limit the parts may stick to the die, while above this the elongation may suffer from the growing $Al_{12}Mn_3Si_2$-phases [4]. Copper and zinc are limited respectively to maxima of 0.03% and 0.07%, leading to a massive improvement in corrosion resistance. A titanium content of between 0.04–0.15% refines the grain structure and improves castability. 100–200 parts per million of strontium are added to modify the eutectic silicon, which turns from unmodified lamellae into a fine coral-like spongeous structure with consequent positive effects on elongation. Magnesium is the 'active' alloying element and its level is set at between 0.1–0.5%, according to the final mechanical properties to be achieved. A low Mg content makes the alloy more ductile with lower yield strengths, whereas a high Mg content favours the yield strength and decreases elongation.

Four major application fields for automotive parts can thus be defined:

1. 0.10% to 0.19% Mg for crash relevant ductile components and for flanged parts, e.g. vibration dampers, housings, spring cups, space-frame nodes, door panels or bumper bars.
2. 0.19% to 0.28% Mg for safety parts under fatigue loads, e.g. engine mountings, flywheel housings or tilting joints. Additionally for T4,T6 and T7 heat-treated parts.
3. 0.28% to 0.35% Mg for components undergoing high operating loads or impact stresses, e.g. structural frames, crankcases, steering columns, bedplates, wheel hubs or spindles.
4. 0.35% to 0.50% Mg for T5 heat-treated castings, e.g. pillars, structural parts, rear suspension arms, engine cradles or strut supports.

Table 6.2 The percentage chemical composition of Silafont®-36

	Si	Fe	Cu	**Mn**	**Mg**	Zn	**Ti**	**Sr**	Others total
Minimum	9.5			**0.5**	**0.1**		**0.04**	**0.010**	
Maximum	11.5	0.15	0.03	**0.8**	**0.5**	0.07	**0.15**	**0.020**	0.10

Alloying elements shown in bold.

The key feature that distinguishes Silafont®-36 from other common HPDC AlSiMg-alloys is its high elongation in the as-cast state. Typical properties of Silafont®-36 are a yield strength of 120–150 MPa coupled with an elongation of 5–12%. The presence of magnesium allows for heat-treating with or without solutionizing of the castings. In recent times, due to its simplicity, efficiency and economical advantages compared to full heat-treatments, the T5 state is preferred as a means of raising yield strength by almost 100 MPa from temper F, while still maintaining ductility. When largest elongations are required, the T4 treatment is performed, whereas T6 produces maximum strength. Overageing to T7 is carried out as a compromise when both properties are required. The wide range of mechanical properties that can be attained by Silafont®-36 in different tempers is summarized in Fig. 6.3.

In addition to static properties, dynamic properties are often important for the design engineer. Recent research has shown that tensile tests performed under dynamic conditions[1] result in significantly higher tensile properties, as shown in Fig. 6.4. In fact, proportionally to strain rates, $R_{p0.2\%}$ increases by 35% to 40%, R_m by 5% to 7% and $A_\%$ by 30% to 35%, regardless of the heat-treated condition. Therefore, in terms of energy absorption, the behaviour of a real casting is improved even compared to static evaluation.

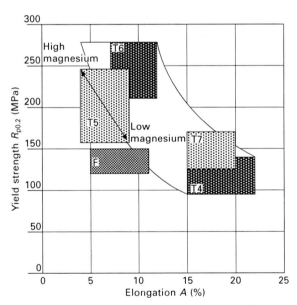

6.3 Yield strength and elongation of Silafont®-36, according to different magnesium contents and heat-treatments.

[1]i.e. traction speed = 6 m/s, which equates to strain rates of around 300 s⁻¹.

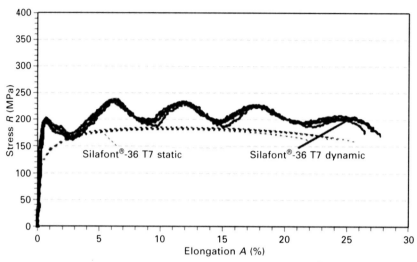

6.4 Dynamic versus static tensile tests for Silafont®-36 in the T7 state.

The fatigue strength is around 90 MPa, assuming a stress ratio of −1 on a 4 mm wall-thick casting in the as-cast state, with a pulsing stress of 117 Hz and a sample geometry of K_t = 1.2. This equates to 66% of the average yield strength of Silafont®-36. As a comparison, a common HPDC alloy AlSi9Cu3(Fe) has a fatigue strength of around 70 MPa. Heat treatment has not been observed to have significant effect on the fatigue life of Silafont®-36: fatigue strength values are almost equivalent to temper F values.

The use of primary low-iron alloys also affects the after-casting operations and finishing steps carried out on components. The improved corrosion resistance allows for uncoated bodywork and car chassis components, while stress corrosion cracking is not an issue.

Traditional jointing technologies are upgraded with new modern technologies. MIG, TIG and spot welding processes can be carried out with good results, as the chemistry of the alloy is free from low-melting impurities or disruptive elements. High-energy fully automatized welding processes have been evaluated on beads-on-plate Silafont®-36 samples for different welding conditions and parameters. Electron beam welding (EBW) processes result in sound welding beams, if a double run is adopted. Laser beam welding (LBW) technology can be applied with fine control of welding parameters, particularly the laser beam energy output and thus the keyhole dimensions. Friction stir welding (FSW) is the most promising modern welding technique that offers positive results on a wide range of welding parameters [5].

Flanging can be adopted by setting a low magnesium content; this offers the great advantage of combining different materials, such as aluminium, steel and plastic. Clinching and self-pierce riveting are jointing processes of growing interest in the automotive field, as they are cheap, fast and do not present the disadvantages of welding seams.

Self-pierce riveting can be performed after a single step heat treatment on Silafont®-36 and is normally preferred to clinching because of the easier access on one single side of the joining area.

The processing of primary low-iron alloys in a HPDCing foundry requires a finer foundry practice [14]. The alloy can offer the best mechanical properties only with the finest tuning of the melting, ladling and casting stages [21]. It is therefore important to avoid pollution by iron, zinc and copper. In most cases, this requires dedicated furnaces, if not a dedicated foundry layout. Magnesium and strontium need to be kept within the defined limits, integrating their content if necessary. The oxides and hydrogen content must be kept to a minimum. This calls for some additional hardware, such as a degassing rotor and a spectrometer with a Sr-channel. The authors consider these devices a must in a modern HPDCing foundry.

The melt temperature should not fall below 630 °C in order to avoid segregations by manganese and should not rise above 780 °C in order to minimize gas pick-up and melt oxidation. Ladling operations should be carried out under laminar flow of the metal. The amount of returns and scraps should be defined on a case-by-case basis according the requirements of the final product. The use of 100% virgin ingots as charges is not exclusive to the automotive industry.

The casting temperature is around 20 °C higher than that of common secondary casting alloys, with consequent higher thermal stress on the foundry equipment. Thermoregulation, the use of conventional die-lubricants at higher concentrations or new developed die-lubes can help to extend the tooling life, including for special casting requirements. Foundry trials showed that preheating the shot sleeve to above 190 °C and the shot piston to 100 °C proves to be the best combination for minimizing presolidifications and inclusions in the casting, while also offering beneficial effects for the elongation.

Clearly, the introduction of premium HPDC alloys does not end with the change of material used, but involves the entire manufacturing process. Analysis of the benefits versus the costs is the key instrument for the decision makers.

The industrial applications of Silafont®-36 are numerous. The first major historical application is given in Fig. 6.5; this use is significant as it marked a turning point in chassis design and construction. The very first Audi surface frame (ASF) of the Audi A8 was designed with 35 high-pressure die-casting nodes. The A-pillar assembly was constructed of two nodes of MIG welded on to an aluminium profile. The main requirements were medium yield strength, weldability, corrosion resistance and most of all ductility, which was required to be above 15%. These demands were fulfilled with vacuum-HPDCing and

6.5 High-pressure die-casting nodes of the A-pillar in the first ASF of Audi A8. Weight: 4.1 kg. Dimensions: ca. 850 × 550 × 90 mm. Temper: T7.

a T7 heat-treatment, which in the mid 1990s was a groundbreaking solution; it suddenly became possible to substitute the former steel construction with an aluminium space-frame 200 kg lighter.

Figure 6.6 shows the bed plate of an sport utility vehicle (SUV). For the first time, a thin-walled oil filter is integrated into an intermediate housing with a large surface in a single HPDCing. The resulting properties are tightness to oil and resistance to dynamic loads, combined with elongation above 6%.

The rear lid of the new BMW Series 5 Gran Turismo is shown in Fig. 6.7. This construction permits the integration of several components and allows for the double opening of the back door. Long-term crash properties,

6.6 Bed plate with integrated filter for a SUV. Weight: 5.8 kg.
Dimensions: ca. 650 × 400 × 155 mm. Temper: T5.

6.7 Rear lid frames of the BMW Series 5 GT. Weight: 11.6 kg.
Dimensions: ca. 1230 × 1250 × 390 mm. Temper: F.

varnishing quality, and weight reduction by increased stiffness were the main requirements. The dimensional accuracy of this 3 mm thick walled casting is impressive. This set of requirements is met by an aluminium casting that allows complicated machining operations and also permits a greater freedom in design than the steel option.

The body structure also benefits from the use of low-iron HPDC alloys. Figure 6.8 shows the integral engine cradle of a middle range passenger car. The adoption of a vacuum-cast Silafont®-36 with 0.28% magnesium and a T5 heat-treatment resulted in 120 MPa yield strength, 250 MPa UTS with 6% elongation.

Fig. 6.9 shows a side door panel of the BMW Z8, whose bodywork is totally manufactured in aluminium. Both side doors are constructed by combining aluminium pressure die castings in Silafont®-36, extruded sections and aluminium sheets. Accurate construction of the door edge and the connection of add-on parts by means of cast sockets are the key aspects of this application.

The engine mounting presented in Fig. 6.10 was developed by BMW for its first hybrid Al-Mg motor block. The engine block in magnesium imposes severe operating conditions on the engine mountings, due to contact corrosion. Thanks to its low Cu, Zn and impurity content, Silafont®-36 offers advantages over other AlSi-pressure die casting alloys.

The side-door window design of the Audi A4 is suitable for welding and provides a very rigid structure which fulfils the maximum manufacturing safety requirements during casting and punching (see Fig. 6.11). The castings cover several functions and are used to improve assembly accuracy, while the structural rigidity is crucial in providing for low wind noise at high speed. Another successful industrial application for vibration dampers is given in Fig. 6.12. The internal rubber-metal elements (Fig. 6.12(a)) are enclosed in the housings by means of flanging (Fig. 6.12(b)). Silafont®-36 with a

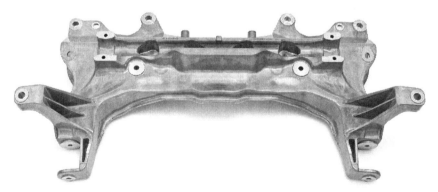

6.8 Integral engine mounting of a middle-class passenger car. Weight: 10.0 kg. Dimensions: ca. 880 × 420 × 350 mm. Temper: T5.

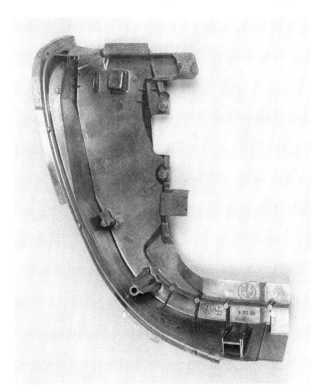

6.9 Side door panel of the BMW Z8. Weight: 1.5 kg. Dimensions: 520 × 440 × 170 mm. Temper F.

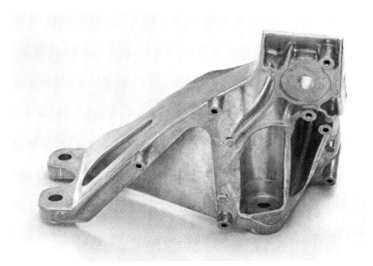

6.10 Engine mounting of a BMW upper-class car. Weight: 1.5 kg. Dimensions: 270 × 170 × 210 mm. Temper F.

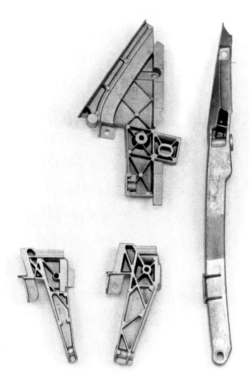

6.11 Side door cast nodes for the Audi A4. Total weight of 10 castings: 3.5 kg. Dimensions: up to 510 mm long. Temper F.

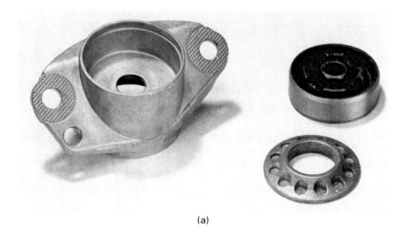

(a)

6.12 (a) Housing for vibration dampers. Weight: 0.2 to 0.4 kg. Dimensions: 80 × Ø 50 mm. Temper F; (b) the assembled damper: Silafont-36 with 0.16% Mg; (c) cross section of the flanged 'crown' of the vibration damper. The elongation values needed on the casting for the performing of this operation is above 8%.

(b)

(c)

6.12 Continued

magnesium content of 0.16% fulfils the high dynamic requirements in the screw-on area as well as the requirement to reduce overloads by ductile deformation during off-roading. This deformation capability is also exploited in the flanging of the closing edge of the housing (Fig. 6.12(c)).

6.3   AlMg non heat-treatable alloys – Magsimal®-59

Magsimal®-59 is a registered trade mark defining a primary casting alloy that chemically corresponds to AlMg5Si2Mn. This alloy was developed and patented by Aluminium Rheinfelden in 1995 and logically and historically followed the concept of Silafont®-36. At that point, an AlMgSi-alloy for HPDCing was unusual; so far this alloy system had been used almost exclusively for permanent mould casting and principally for sea-water applications. The main reason behind the design of this new material was the desire to combine the superior mechanical properties of the AlMg-system with its application in HPDCing purely in the as-cast state.

The nominal chemical composition of Magsimal®-59 is given in Table 6.3. The main alloying elements are magnesium, silicon and manganese. Magnesium goes into solution and results in high yield strength by forming coherent and semi-coherent phases in the α-aluminium phase. The ratio of magnesium to silicon is very important in obtaining the desired share of 40–50% eutectic. This in turn favours sufficient castability and feeding during solidification. No free silicon must be available in order to provide outstanding corrosion behaviour. Excess magnesium forms Mg_2Si compounds – to be seen in the typical ternary eutectic microstructure. Again, as for Silafont®-36, manganese prevents the material from sticking to the die and lowers the solubility of iron into the aluminium, thus substituting the function of iron itself [6] [7]. Manganese is responsible for forming Al_6Mn intermetallics, whose morphology is formed of irregular polygons.

The iron content is kept below 0.2% in order to minimize the formation of the needle-like β-AlFeSi phases, whose morphology would initiate cracks in the casting under load, deteriorating tensile strength, elongation and fatigue strength. Titanium is normally used for grain-refining purposes. Metallographic studies have shown, however, that a grain refinement with titanium/boron has a deteriorating effect on the fine morphology of the ternary eutectic of Magsimal®-59. For this reason, no TiB_2-particles should be added. Magsimal®-59 has a special long-term grain refinement performed during manufacturing and targeting the $AlMg_2Si$-eutectic. Around 240–250 ppm of vanadium and zircon are additionally alloyed in order to achieve further grain refinements in the structure and to improve the tensile properties.

AlMg-melts with an Mg content of over 2% have a tendency towards an increased dross formation, especially if the melt is maintained for a long

Table 6.3 The percentage chemical composition of Magsimal®-59

	Si	Fe	Cu	**Mn**	**Mg**	Zn	Ti	**Be**	Others total
Minimum	1.8			0.5	5.0				
Maximum	2.6	0.2	0.05	0.8	6.0	0.07	0.20	0.004	0.2

Alloying elements shown in bold.

period in holding furnaces at elevated temperatures [8]. Beryllium has the capacity to increase the density of the superficial oxide skin, forming a BeO film on its surface [9]. Alloying 30–40 ppm Be during manufacturing allows the resistance of Magsimal®-59 melts to be improved, as less aluminium and magnesium diffuse to the external surface.

The alloy is produced on the basis of a 99.8% pure aluminium. The control of impurities and minor elements is therefore guaranteed. Copper and zinc are kept below the maximum levels of 0.05% and 0.07% respectively. This complies with automotive standards and provides outstanding corrosion resistance. Calcium and sodium are limited to a maximum of 15 ppm, as these enhance the hot-tearing tendency of the castings and lead to poor fluid flow.

Contamination by phosphor must be avoided at all costs, as its influence on the eutectic fineness is damaging even in the range of 10–15 ppm. According to internal research, the presence of few parts per million of As, Ba, Te, Hg, Be, and K also has the same negative effect.

This very fine ternary eutectic is the special feature of Magsimal®-59 and results in high mechanical properties and ductility in the as-cast state. The microstructure is achieved by a special manufacturing process, which transforms the Mg_2Si eutectic from its normal coarse brittle plate-like structure into a more ductile fine fibrous structure. The micrographs shown in Fig. 6.13(a) and (b) clearly depict the huge difference between Magsimal®-59 and a common HPDC AlMg5Si2Mn alloy.

The mechanical properties of AlMg-HPDC alloys, such as Magsimal®-59, show a more pronounced decay due to increasing wall thicknesses in comparison to AlSi-HPDC alloys [15] [17]. This is a boundary condition that the experienced designer has to take into account. The 'value-added' benefit of Magsimal®-59 can be considered to be the delivery of superior mechanical properties in the as-cast state, i.e. after casting without any further heat treatment, as shown in Table 6.4.

A curious feature typical of AlMg-alloys is the so-called 'strain-induced ageing' of the material during tensile testing. This phenomenon occurs in the plastic area of the stress–strain curve and is shown by a large number of small irregularly distributed peaks. From an atomic point of view, this is an interaction between solid solution atoms and migratory dislocations in the microstructure, which causes a momentary low stress reduction in the tensile curve itself. The phenomenon is also known as the 'Portevin – Le Chatelier effect' and is clearly visible in Fig. 6.14.

Full heat treatments are not normally carried out with Magsimal®-59 for two main reasons:

1. the as-cast properties of the alloy approach the performance of common AlSi-alloys in the T6 state and

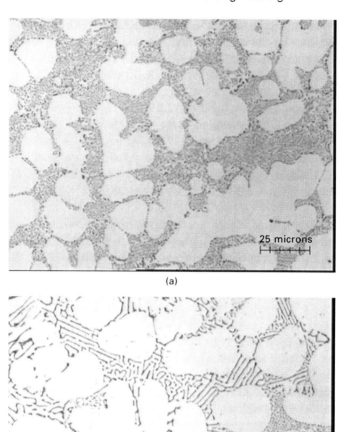

(a)

(b)

6.13 The microstructure of Magsimal®-59 – (a) compared to a common AlMg5Si2Mn alloy - (b). The different morphology of eutectic is the key feature for superior elongation and fatigue strength.

2. full heat-treatments do not bring about substantial improvements in the mechanical properties in AlMgSi systems [10] [20].

Additionally, all the negatives connected to full heat treatments of high-pressure die-castings are avoided. However, heat treatment can be carried out on Magsimal®-59, according to the following modalities:

Table 6.4 Mechanical properties of Magsimal®-59 in the as-cast state, as a function of the wall-thickness

Wall thickness (mm)	$R_{p0.2}$ (MPa)	R_m (MPa)	A (%)
< 2	>220	>300	10–15
2–4	160–220	310–340	12–18
4–6	140–170	250–320	9–14
6–12	120–145	220–260	8–12

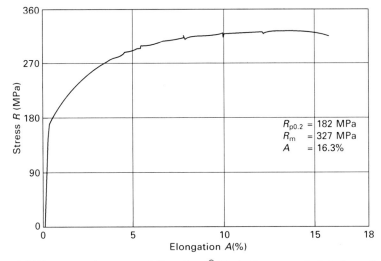

6.14 Stress-strain curve of Magsimal®-59, in the as-cast state for a 3 mm-wall thick sample.

- T5 – to increase the yield strength
- O – to increase the elongation.

T5 consists of quenching the castings in water directly after their removal from the die and subsequently artificially ageing them to a well-defined temperature and time. It is of a vital importance that the castings should be quenched immediately after removal from the die, in order to 'freeze' the soluted magnesium into the matrix, preventing the subsequent hardening. Air cooling is not effective and does not result in the desired mechanical properties after ageing.

Stable outcomes by age-hardening can be observed after approximately 60 minutes treatment. 200–250 MPa can be targeted for 3mm-thick castings. Yield strengths of over 200 MPa can be obtained on thicknesses of 6 mm. Elongation suffers as a result of a T5 treatment and the values fall to an average of 10% and 4% respectively for 3 mm and 6 mm wall thickness, after 60 minutes ageing time.

In summary, a T5 heat treatment can increase the yield strength by up to 30%, while reducing elongation by up to 30%. The so-called 'O' treatment consists of a simple annealing at low temperature, commonly set between 320 °C and 380 °C. Typical annealing times are 30–180 minutes. Two main kinds of O tempers can be defined:

- O (I) with annealing at lower temperature, i.e. around 320 °C
- O (II) with annealing at lower temperature, i.e. around 380 °C.

As annealing at low temperatures does not have the same strong impact as a T5 treatment on the mechanical properties of Magsimal®-59, O temper is normally intended as O(II). Different annealing temperatures have been investigated; higher ductility can easily be easily achieved by treatment at 350 °C for 60 minutes. In summary, the O heat treatment can cause yield strength to decrease by up to 30%, while increasing elongation by up to 30%. The choice of the correct treatment is made after evaluation of the requirements set by the designers. The definition of the correct treatment parameters takes place as fine tuning, *ad hoc* on the casting geometry. Stabilizing castings is recommended if they are required to work under thermal stress. This can be achieved simply by heating the castings to slightly above the working temperature for 2–3 hours.

The properties of a material under dynamic load or multi-axial stresses are a fundamental criterion for the design engineer. In fact, they are a more precise reflection of the behaviour of castings during their operating life. Fatigue performances are a direct function of the chemistry and solidification conditions of the metal. However, surface conditions and metallurgical defects influence the fatigue strength to a huge extent. Measurements carried out with a high-frequency pulse generator (approx. 110 Hz) and a 200 kN load give the Wöhler's curves traced in Fig. 6.15 for 5%, 50% and 95% fracture probability. If the 5% fracture probability curve and the value of 10^6 cycles are considered, as is common practice, Magsimal®-59 shows a fatigue strength of 100 MPa in the as-cast state.

This value is significantly higher than that of the AlSi alloys. In fact, under the same loading conditions an AlSi7Mg0.3 (A356), cast in permanent mould and heat-treated to the T6 state, has a fatigue limit of 93 MPa. EN 1706:1998 normed HPDC alloys show fatigue limits of 60–90 MPa.

Tensile tests under dynamic conditions have shown that the yield strength increases by approximately 16%, i.e. to a lesser extent compared to AlSi alloys. R_m and $A\%$ improve by about 5% and 27% respectively compared to static tensile tests.

Generally speaking, AlMg alloys are the best choice in terms of corrosion behaviour. This also explains why they are used in sea-water applications. The adoption of AlMg systems for safety components also immediately raises the issue of their tendency towards stress corrosion cracking. In fact,

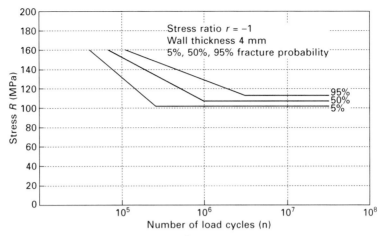

6.15 Wöhler's curve for Magsimal®-59.

pure AlMg alloys with an Mg content of over 3% suffer from both SCC and intercrystalline corrosion. This is due to Mg_2Al_3 phases that precipitate at the grain boundaries if the temperature increases up to 60–150 °C. These phases are not precipitated by Magsimal®-59. Tests performed with samples in Ma-59 according the ASTM G 47-90 confirmed that SCC was not an issue with this material [18]. The combination of a corrosive atmosphere with dynamic loads constitutes subjecting a material to its severest operating conditions. This combination, however, is typical of automotive applications, such as crossbeams, engine cradles or integral mountings. Figure 6.16 shows how pressure die-castings in Magsimal®-59 can compete with thixo-castings in common AlSi7Mg0.3 T6 and with fully heat-treated sheet metals of 6000 series. This can be explained by the chemistry and the uniform microstructure of Magsimal®-59, which does not present any precipitates at the grain boundaries, where corrosion can easily be initiated.

If primary casting alloys are ranked according to their ability to be welded, AlSi alloys can be considered superior to AlMg alloys. Nevertheless, MIG and TIG structural welding with Magsimal®-59 can be carried out by using metal and AlMg4.5Mn master alloy as filler [19]. As is well known, elongation values are affected to a large extent in the HAZ after welding, if compared to the static strength properties (see Table 6.5). Experiences with Magsimal®-59 show that elongation drops even more if the filler metal is a common AlSi5. For this reason, welded seams are normally positioned in low loaded areas by the casting designer; they should also be located not far from the ingate, taking into account the nature of the high-pressure die-casting process.

Self-piercing riveting can be used with Magsimal®-59, if the casting is not thicker than 3 mm, in order to guarantee higher elongation and in turn the required wall deformability. A combination of self-piercing riveting

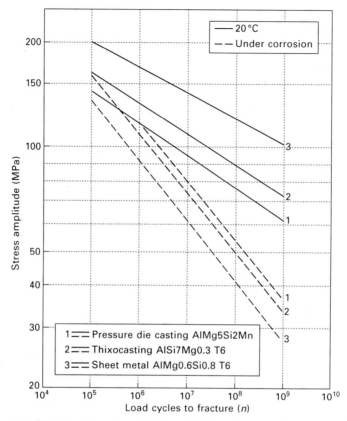

6.16 Fatigue properties of different alloys in corrosive context. Curve 1 depicts the behaviour of Magsimal®-59 (source: Halden wanger, Seminar Werkstoffvielfalt im Automobil, February 2000).

Table 6.5 Tensile tests performed on welded and non-welded samples in Magsimal®-59. Manual MIG welding with AlMg4.5Mn filler metal

Wall thickness (4 mm)	$R_{p0.2}$ (MPa)	R_m (MPa)	A (%)
Unwelded	165	287	17
MIG-welded	148	246	6

and structural gluing is compatible with the alloy and is today considered a common jointing method.

Castings in Magsimal®-59 can be easily painted or powder coated. Cataphoretical painting is the most common technology in use today in the automotive industry.

Polishing and protective anodizing can be carried out with relatively good results. Polishing produces a typical light blue colour on the surface gloss. The surface should finally be protected against scratches by a transparent varnish. Gray-shaded zones are visible after anodizing, due to the presence of silicon. Glass blasting is recommended in order to balance the heterogeneous appearance of the surface. If the oxidation layer only serves a protective purpose, no further technical issues need to be taken into account. The aesthetics of Magsimal®-59 after anodization are preferable to those of AlSi alloys; however, chromium coatings are recommended for decorative purposes.

Magsimal®-59 seems a 'difficult' alloy to deal with as its physical properties are significantly different to those of the commonly used AlSi alloys. Due to its average silicon content of 2%, the alloy has higher volumetric shrinkage compared to AlSi alloys; reference values are respectively 0.6–1.1% versus 0.4–0.6%. Therefore, some guidelines should be kept in mind when Magsimal®-59 in used in design. Ribs should not be designed to be too thin, as the alloy delivers its best performances in thin sections. 1–2 mm ribs connected to 6 mm walls are not recommended as the whole construction would have an extremely undesirable rigidity. Additionally, deformation under stress would be located at the walls of the castings, and not on the ribs. Nodal points with massive material presence should be avoided. A sinking of the surface could be observed, because of the higher volume contraction. Material agglomerations on internal radii can also cause sinking points. The adoption of 'crow's feet' provides a solution to this problem. In unfavourable solidification conditions and for certain castings, shrinkages can end in a central line, normally located on the central neutral fibres of the material itself. These have no evident effect on the strength of the casting. However, it is extremely important in this instance that the shrinkage does not come into contact with the open surface of the heavy-duty section of the casting.

Wide casting surfaces with high requirements should be designed with the main criterion of the laminar flow of the metal. Openings and drillings should be machined after casting and not obtained as cast holes. Regular wall-sections should be evaluated together with uniform temperature distribution on the die. If hot spots are present, a premature crack of the casting could occur. Draft angles should be generous, above 1.5°, and preferably 2°. The tooling and die equipment used is one of the most important factors contributing to the total cost of a casting.

Magsimal®-59 is a primary alloy with a maximum iron content of about 0.2% and a eutectic temperature of around 30°C higher than that of AlSi10Mg(Fe). These factors must be taken into consideration during die manufacture. Die lubricants should be used at 30–50% higher concentration than used for traditional AlSi alloys. The most recent generation of die lubricants developed for Mg alloys can help to protect the die for a longer period at higher temperatures, while favouring the flow of metal into the die cavity. Generally speaking, dissipating heat from the die just by means of die lubrication, i.e. from the die surface, shortens the life of the die. Water-cooled or oil-cooled tools with heat-exchangers, on the other hand, improve the life of the die.

It is clear, then, that the special chemistry of Magsimal®-59 needs to be managed in the correct way. This is turn requires a precise foundry practice, even stricter than required for Silafont®-36. Dedicated crucibles should be used to avoid contamination by impurities. Quartz-free refractories with a high content of aluminina – Al_2O_3 content are preferred. Ingots should be melted quickly to avoid gas pick-up, magnesium and melt oxidation, possibly in gas fired furnaces. Convection helps the homogeneous mixing into the melt. The temperature after melting should not exceed 780 °C and should not fall below 650 °C during holding, to avoid segregations by manganese. No degassing tablets, modification or grain-refining additives should be used. The expected loss of magnesium is normally between 0.10% and 0.15%, depending on melting conditions. Internal experiments showed that a rectification of the Mg-content should be avoided. Addition of pure magnesium can be carried out, up to 0.5%; beyond this value, the microstructure will be coarsened. For the same reason, the maximum scraps and returns should not exceed 50%. A 30% recycling rate is the suggested value for series production. Only ingots should be used, if highest elongation values and fatigue strength are the key requirements for the final casting. The aluminium share in the dross can be reduced by using appropriate salts developed for Magsimal®-59. Covering the melt delivers drier dross than mixing the salts into the melt. Nitrogen and/or argon rotor degassing is considered a common practice in a modern foundry. This operation should be intensified when scraps and returns are present.

Applicative examples of series productions using Magsimal®-59 clearly show the application should fit the nature of this alloy in order to fully exploit its potential.

Light-design is shown in Fig. 6.17, which presents the inner door panels of an SUV. The length of the parts is around 1.2 m with an average wall thickness of 2 mm. The required ductility is provided in the as-cast state, avoiding the need for heat treatment and therefore the risk of distortion. By designing all four doors this way, the vehicle weight was reduced by 40 kg compared to the previous steel construction, while a better ratio of rigidity to weight was also provided. Further advances led to other parts being developed in the same way; these can also be welded.

A smart solution for a passenger car has been adopted by casting one frame door by one HPDCing shot, as shown in Fig. 6.18. The whole construction has excellent rigidity, with a substantial weight saving, wisely distributed from the bottom to the top of the door. Laser trimming has been implemented in the HPDCing process and further exploited to cut off the openings in the upper area of the part, thus taking lightweight construction to its limits.

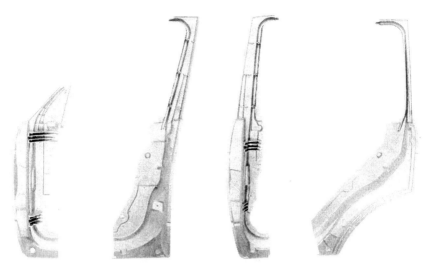

6.17 Inner door panels of a SUV. Weight: 2.0–2.2 kg. Dimensions: ca. 1000 × 240 mm to 1400 × 500 mm. Wall thickness: 2–2.5 mm. Temper: F.

6.18 Door frame. Weight: 3.6 kg. Dimensions: ca. 1110 × 852 × 261 mm. Temper: F.

In order to spread the weight of the vehicle evenly between the front and back wheels, the frontal subframe is today commonly designed in aluminium. High static loads combined with alternating loads are the operating conditions for the shock tower shown in Fig. 6.19. The shock tower is connected to the

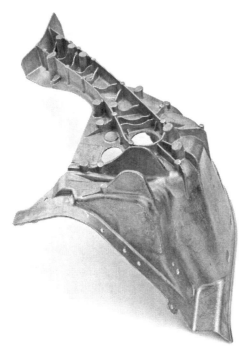

6.19 Shock tower. Weight: 2.3 kg. Dimensions: ca. 500 × 280 × 500 mm. Temper: F.

suspension strut and must therefore withstand the static load imposed by the weight of the car and the dynamic shocks imposed by the suspension. Thin walls provide a high yield strength and elongation. High fatigue strength and toughness are also required. This weldable part is connected to the surrounding profiles by self-pierce riveting.

The integral cradle presented in Fig. 6.20 has been designed to work in the as-cast state. A heat treatment would have caused a distortion of the two arms, which is incompatible with the requirement for high dimensional stability. Aluminium profiles are welded on to the part at several points, in order to allow connection to the rest of the chassis.

High dynamic loads under high static load with corrosion resistance in the under-body area were characteristic requirements for the gearbox crossbeam of the Mercedes-Benz S Class (see Fig. 6.21).

Vibration dampers are pressed into the two machined bores, from where the load is discharged into the plate. Despite these high loads, the wall thickness could be kept between 4–5 mm, as an increase in yield strength to over 190 MPa is achieved by a subsequent heat treatment at 200 °C.

Steering-wheel frames must not break in the event of a crash, but should only distort behind the steering column plate without forming sharp edges

6.20 Integral crossbeam for BMW Allroad, Series 5, 6 and 7. Magsimal®-59. Weight: 4.1 kg. Dimensions: ca. 810 × 450 × 210 mm. Temper: F.

6.21 Gearbox crossbeams of the Mercedes S-Class. Weight: 2.3 kg. Dimensions: 610 × 210 × 75 mm. Temper T5.

(see Fig. 6.22) They are subjected to strict qualification tests, e.g. an impact test with a dummy in the front seat or dimensional stability control at high temperatures that simulate desert climates.

The suspension strut bracket of Fig. 6.23 has been used in several SUV models. This component is subject to high dynamic loads and is therefore designed in Magsimal®-59 with wall thicknesses of 4–6 mm. This design is without thicker nodal points or thicker sections only in low loaded positions, e.g. on the double T cross-section, shows that this application is a suitable use of Magsimal® 59.

6.22 Steering wheel of the VW Beetle. Weight: 0.85 kg. Dimensions: Ø 350 × 125 mm. Temper F.

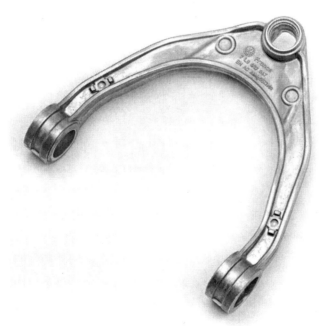

6.23 Suspension strut bracket for SUVs. Weight: 0.9 kg. Dimensions: 340 × 370 × 60 mm. Wall thickness: max. 6 mm. Temper F.

6.4   AlSi non heat-treatable alloys – Castasil®-37

The increasing design complexity of automotive components and the continuous growth in their dimensions make the use of HPDC AlMg alloys very difficult due to their poorer flow abilities. Therefore, these parts are mainly cast in AlSiMg alloys with subsequent heat treatment to achieve the required properties. However, if solutionizing is carried out, the risk of blistering and distortion of the casting is high. This obstacle opened the way for a new field of application. As in body structures, the properties most frequently required are medium yields – above 120 MPa – and high elongation – above 12%; an alloy capable of complex designs for structural parts in the as-cast state makes the heat treatment obsolete.

With these aims in mind, a new alloy was developed and introduced into the market in 2003 under the trade name Castasil®-37. It combines the good castability of the aluminium–silicon system with the above-mentioned requirements. Additionally, no age hardening potential is made available. In fact, since engine specific powers and hence operating temperatures have increased in recent years, structural components mounted around the engine compartment are now exposed to a warmer environment. No long-term ageing is allowed as the castings must have stable properties throughout their service lifetimes. The chemistry of Castasil®-37 is given in Table 6.6.

As with Silafont®-36, 99.8% pure primary aluminium is the basis of the metal. An average silicon content of 10% provides excellent castability and, since it expands during solidification, volumetric shrinkage is comparable to that of other common AlSi alloys. The copper and zinc contents are limited to 0.03% and 0.07% respectively for corrosion reasons. The iron content is kept below 0.15% to prevent the formation of needle-like AlFeSi phases, while Mn is present to improve the die ejection of the castings.

Foundry experiments confirmed that the optimum Mn level is between 0.35 and 0.60%. The most crucial difference between Castasil®-37 and Silafont®-36 is the lower magnesium content of the former. Research trials for both long- and short-term ageing have shown that the threshold of magnesium content for preventing significant age-hardening phenomena is below 0.06% [11]. However, if even higher elongation is required, a single-step heat treatment to temper O can be performed in order to obtain the mechanical properties listed in Figure 6.24.

Table 6.6 The percentage chemical composition of Castasil®-37

	Si	Fe	Cu	**Mn**	Mg	Zn	**Mo**	**Zr**	Ti	**Sr**	Others total
Minimum	8.5			0.35						0.006	
Maximum	10.5	0.15	0.05	0.60	0.06	0.07	0.3	0.3	0.15	0.025	0.10

Alloying elements in bold.

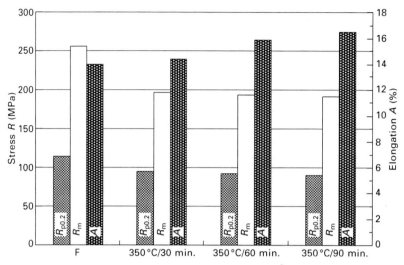

6.24 The mechanical properties of Castasil®-37 in temper F and for different O heat-treated states.

Limiting the magnesium content to 0.06% results in an even finer eutectic structure, as shown in Fig. 6.25, which compares the eutectics of Silafont®-36 and Castasil®-37. Permanent modification is carried out by means of 60 ppm to 250 ppm strontium. Up to 0.03% molybdenum and zirconium is added in order to strengthen the metal, while at the same time providing a grain refining effect. The combined action of both elements proves beneficial to elongation, with improvements of almost 30% compared to those achieved through single additions of Mo or Zr.

Tensile tests performed in static conditions[2] give average mechanical properties of around 120 MPa $R_{p0.2\%}$, 265 MPa R_m and 15.9% elongation; the same tests performed under dynamic conditions result in around 170 MPa $R_{p0.2\%}$, 280 MPa R_m and 19.6%, corresponding to percentual increases of approx. 41%, 6% and 23% respectively (see Fig. 6.26). As with Silafont®-36, a cast part in Castasil®-37 is expected to 'work' better in terms of energy absorption if stressed in a dynamic way, e.g. in the event of a crash.

Similar conclusions have been drawn by the Swiss company Bühler in their testing of real components [12]. A qualitative test carried out with a 2 kg steel ball dropped from a height of 2 meters on a 2 mm casting in Castasil®-37 showed that the alloy deforms without cracking. The same test on a common 226 alloy shows only minor deformations, but in most cases incipient cracks are observed. Compression tests performed on 2 mm castings give the results shown in Fig. 6.27. A steel punch deforms the casting until

---

[2]i.e. traction speed = 0.02 mm/s which equates to 4–6 $*10^{-4}$ s^{-1}.

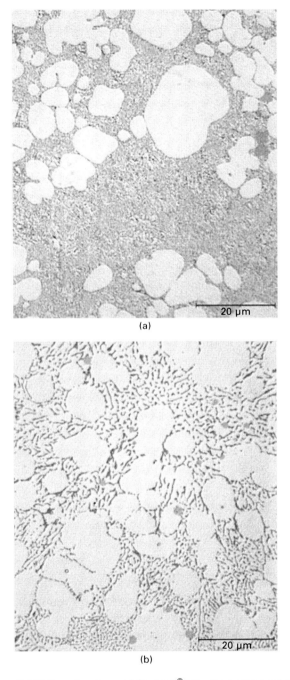

(a)

(b)

6.25 Microstructure of Castasil®-37 - (a) compared to Silafont®-36's one – (b) As-cast state, 3 mm wall-thick sample.

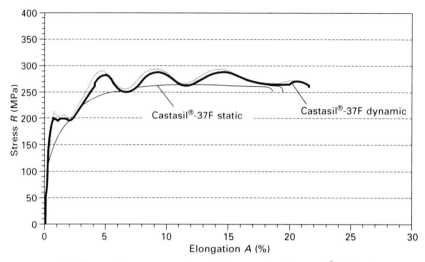

6.26 Dynamic versus static tensile tests for Castasil®-37 in the as-cast state.

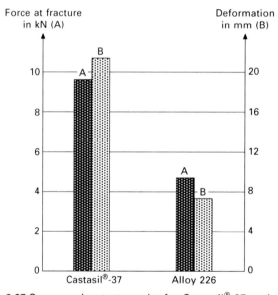

6.27 Compression test results for Castasil®-37 and a common 226.

fracture. The path of deformation and maximum force are measured, using methods similar to the 'Erichsen test' which is normally carried out on metal sheets. Deformations are almost 3 times higher with Castasil®-37 than with a common 226 alloy, while borne loads are also doubled.

The fatigue strength of the material is around 86 MPa, assuming a stress ratio of −1 on a 4 mm wall-thick casting in the as-cast state, with pulsing stress of 117 Hz and a sample geometry K_t = 1.2 at 5% fracture probability. Wöhler's curves at 50% life probability give a fatigue resistance of 103 MPa.

WIG and MIG welding tests have shown that the alloy itself is weldable. Free-magnesium filler metals AlSi5 or AlSi12 should ideally be used. The soundness of the welding seams depends to a significant extent on the quality of the HPDCing process, but virtually castings have been obtained that are virtually free of pores [12].

LBW proves to be more effective with Castasil®-37 than with Silafont®-36, as the latter contains magnesium, which sublimates during the welding process and thus favours gas formation. Again, the control of the laser power is the key variable for sound seams; slow welding cycles allow a longer permanence at the liquid state of the alloy, with a self-degassing effect. Round screwed pin FSW experiments on Castasil®-37 gave excellent results in terms of quality of the welding seam; in comparison to Silafont®-36 higher welding speeds can be adopted of up to 16 mm/s on 4 mm wall thicknesses. Sound EBW can be carried out with welding speeds of between 20 and 40 mm/s on wall thicknesses between 2 to 4mm, typical of HPDCing geometries. The second run and an optional third cosmetic run lead to the best results in terms of seam quality [5].

Due to the absence of magnesium the ability of Castasil®-37 to be flanged is better than that of Silafont®-36. To achieve crack-free flanged parts, the metal should elongate by at least 8% which requires a correct configuration of the flanging edges and a careful design of the gating system in order to grant good metal flow in the flanging area. These requirements are also valid for flanging through cone rolling. Self-pierce riveting, where the casting is placed as the lower layer, place the highest demands on the quality of the casting itself. Industrial applications and experiences confirmed that Castasil®-37 can be self-riveted in temper F under strict design conditions, i.e. just with a rivet-die of flat geometry. Otherwise a further improvement in deformability is achieved in the temper O, e.g. after 90 minutes at 330 °C.

The management of this kind of alloy in the foundry requires the same precautions adopted for Silafont®-36. Additionally, in order to ensure that the material is not sensitive to age-hardening, pollution by magnesium must be avoided. When melting pre-modified alloys in crucible or furnaces, more strontium will be lost during the first melting procedure as it diffuses into the refractory materials and saturates them. With a second melting batch, equilibrium will be reached. Positive experiences in the field of ductile and weldable HPDCing in Castasil®-37 have been obtained through the 'Structural' process developed in Switzerland. This is essentially based on a standard casting process, outfitted with adequate melt treatment, a special dosing and spraying method and an air/gas die evacuation system. Qualitatively, the Structural casting technique produces the same result as the Vacural method.

Typical applications for Castasil®-37 are wide-surfaced structural parts with complex design and geometries, where high ductility in the as-cast state is needed. Industrial series applications are discussed below with some short notes underlining the features most relevant to the adoption of this premium low-iron alloy.

Figure 6.28 shows some HPDC nodes of the Lamborghini Gallardo Spyder ASF, which give rigidity to the whole construction and allow for the

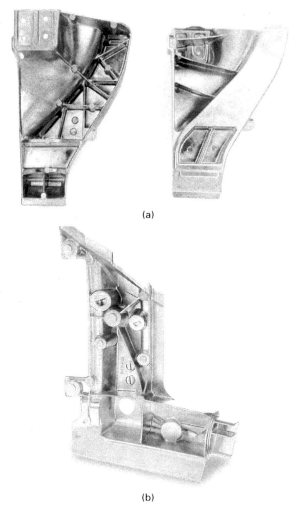

(a)

(b)

6.28 HPDC nodes of the Lamborghini Gallardo Spyder. Weight: 0.6 to 2.0 kg. Dimensions: ca. 210 × 200 × 150 mm to 340 × 210 × 200 mm. Temper: F.

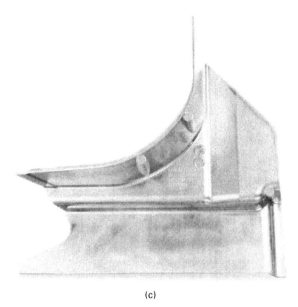

(c)

6.28 Continued

connection of extruded profiles and sheet-metal panels by MIG-welding and riveting.

Figure 6.29 shows the hinge and latch doors panels of the Jaguar XK which has a bodywork entirely in aluminium. The panels are combined with extruded bars and metal-sheets. Stiffness and dimensional accuracy are a valuable additional property offered by the aluminium solution. Elongation above 7% has been reached in the as-cast state, while making the construction 5–8 kg lighter than the steel version.

Figure 6.30 shows the rear connector sill frame member of the new Audi A8. The component, mounted on the chassis in temper F has impressive dimensions: 1.45 m length with a total weight of 10.2 kg. The size and the possibility of casting complex geometries allow the integration of many functions while reducing part numbers. In fact, while the steel version would have been constructed of 20 parts, the previous aluminium version required six additional castings. At the same time, the balance of weight has been achieved by the locally optimized ribbing defined on the mapping of loads.

The cross brace of Fig. 6.31 is assembled directly in front of the dashboard support and increases the stiffness of the flat-designed sports car. Corrosion resistance and high deformation capability were required in the intermediate area between the engine bonnet and the windscreen.

In the event of a crash the hinged levers of the retractable hard top (RTH) are particularly close to the passengers in the vehicle and are therefore subject

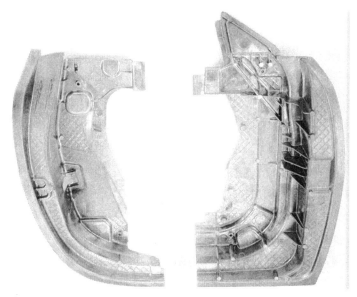

6.29 Hinge and latch door panels of the Jaguar XK. Weight: 1.2 to 2.1 kg. Dimensions: ca. 700 × 340 × 170 mm. Temper: O.

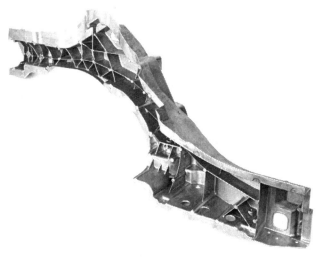

6.30 Rear connector sill frame member of the new A8. Weight 9.902 kg Dimensions: ca. 1454 × 375 × 552 mm. Temper: F.

to the highest ductility requirements. These components, presented in Fig. 6.32, must prevent breaking off in case of an accident. Castasil®-37 fulfils the particular requirements of the VW EOS with an elongation above 7% in the as-cast state.

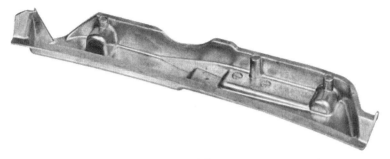

6.31 Frontal cross brace of the Audi R8. Weight: 0.18 kg. Dimensions: 370 × 70 × 60. mm. Temper F.

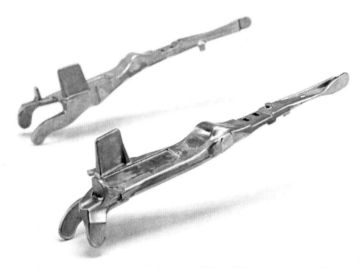

6.32 RHT folding levers of the VW EOS. Weight: 0.56 kg. Dimensions: 510 × 100 × 80 mm. Temper F.

6.5   Automotive trends in die-casting

All the alloys discussed above constitute useful alternatives for automotive components. While Castasil®-37 is best suited to applications in the body structure that require a large elongation rate and moderate yield strengths, Magsimal®-59 and Silafont®-36 are applied principally in the suspension area, where higher yields are required. Temper F can be used only with Magsimal®-59; if heat-treatment parameters are adjusted, Silafont®-36 offers the widest spectrum of mechanical properties.

These alloys can be considered the current state-of-the art in modern HPDCing. The requirements for automotive parts are continuing to follow the trends mentioned in Section 6.1 and it is very probable that these materials will be brought close to their physical limits very soon.

The introduction of low-primary HPDC alloys has substantially enhanced the level of achievable material performance.

The authors maintain that a turning point of this sort will not occur again. AlSi alloys have been chemically fully optimized and improvements are only possible in terms of heat-treatments, e.g. with SST® or similar [13].

There is still some potential for improvement in the chemistry of AlMg alloys. In fact, foundry trials with HPDC AlMg alloys with less than 0.3% silicon showed the possibility of reaching elongations over 20% while maintaining yield strengths above 170 MPa in the as-cast state. The drawback is the absence of silicon, which makes the castability of the metal very poor and not suitable for HPDCing.

HPDC AlZnSiMg alloys are not considered an alternative for automotive requirements, as waiting 10–30 days for the self-ageing of the castings and introducing a protective coating on the parts to prevent corrosion by zinc is not an economical option. Further improvements by means of simple alloying no longer appear to be feasible.

Composite alloys are promising aluminium materials, though they unfortunately do not easily create a homogeneous matrix and reinforcement distribution. Ductility is always hindered. Additionally, metal matrix composite(s) (MMCs) have not found wide application in the automotive industry as the cost of raw materials is very high and the stability of the casting process is uncertain.

Aluminium nanomaterials are the current research trend. However, there is still no evidence of successful alloys that are available for HPDC processing. Apart from the high manufacturing cost of nanoparticles, the main issues are connected to their poor wettability by aluminium and to their toxicity, especially for carbon containing nano-reinforcements.

6.6   References

[1] Aluminium Rheinfelden GmbH, Publication Code 807, *Neue Entwicklungen auf dem Gebiet der warmfesten Aluminium-Gusswerkstoffe, AlCu₅Mn(Ni, Ce), AlMg₂Si(Sc,Zr)*, Sonderdruck aus Giesserei 91, pages 32–38, 08/2004.

[2] M. Garat, G. Laslaz, *Improved Aluminium Alloys for Common Rail Diesel Cylinder Heads*, AFS Transactions 2007, Paper 07-002(02), 2007.

[3] Aluminium Rheinfelden GmbH, *Primary aluminium alloys for pressure die casting*, 1 Edition, June 2007.

[4] Aluminium Rheinfelden GmbH, Publication Code 637, *Optimizing the Manganese and Magnesium Content for Structural Part Application, Silafont-36, AlSi₉MgMn*, 22nd International Die Casting Congress and Exposition by NADCA, Indianapolis, 2003.

[5] G. Luvara, *Studio di Saldabilitá (LBW-FSW-EBW) di Leghe da Pressocolata a Base di Alluminio*, Universitá degli Studi di Genova, Facoltá di Scienze M.F.N., A.A. 2008–2009, Genova, 2009.

[6] Aluminium Rheinfelden GmbH, *Neuentwickelte Druckgusslegierung mit*

ausgezeichneten mechanischen Eigenschaften im Gußzustand, Magsimal-59, AlMg₅Si₂Mn, Sonderdruck aus Giesserei Ausgabe 3/98.

[7] Aluminium Rheinfelden GmbH, Publication Code 632, *Erfahrung aus der Serienproduktion von druckgegossenen Lenkradskeletten in der Legierung Magsimal-59, AlMg₅Si₂Mn,* TMS Annual Meeting and Exhibition, San Diego, March 1999.

[8] Aluminium Rheinfelden GmbH, *Aluminium Druckguss Legierungen,* 1 Edition, May 2003.

[9] O. Ozdemir, J. E. Gruzleski, R. A. L. Drew, *Effect of Low-levels of Strontium on the Oxidation Behaviour of Selected Molten Aluminium-magnesium Alloys,* Springer Science+Business Media, LLC, 2009.

[10] Aluminium Rheinfelden GmbH, *Primary Aluminium Casting Alloys,* February 2006.

[11] Aluminium Rheinfelden GmbH, Publication Code 806, *Non Aging Ductile Pressure Die Casting Alloys for Car Construction, Castasil®-37, AlSi₉Mn,* 08/2004.

[12] Aluminium Rheinfelden GmbH, Publication Code 638, *Economic Production of Ductile and Weldable Aluminium Castings, Castasil®-37, AlSi₉Mn,* Reprint from *Casting Plant and Technology International,* pages 22–27, February 2006.

[13] K. Greven, D. Dragulin, *Ductile High Pressure Die Casting – Heat Treated or Temper F?,* Proceedings of the 2nd International Light Metals Technology Conference 2005, 2005.

[14] Rheinfelden Alloys GmbH & Co. KG, *Hüttenaluminium-Gusslegierungen,* Ausgabe 7, 01/2010.

[15] Aluminium Rheinfelden GmbH, Publication Code 635, *Möglichkeiten des Aluminiumdruckgiessens. Anwendung dieser Technologie im Grenzenbereich, Magsimal®-59, AlMg₅Si₂Mn,* Sonderdruck aus Giesserei 90, Nr. 7, 2003.

[16] Aluminium Rheinfelden GmbH, Publication Code 630, *Producing Low-iron Ductile Aluminium Die Castings,* 18th International Die Casting Congress and Exposition by NADCA, Indianapolis, 1995.

[17] H. Koch, U. Sternau, H. Sternau, A. J. Franke, *Magsimal-59, an AlMgMnSi-Type Squeeze-Casting Alloy Designed for Temper F,* TMS Annual Meeting, Anaheim, LA, February 1996.

[18] Aluminium Rheinfelden GmbH, Publication Code 633, *Experience of Three Years Producing Low Iron Ductile Pressure Die Castings,* 19th International Die Casting Congress and Exposition by NADCA, Minneapolis, 1997.

[19] Aluminium Rheinfelden GmbH, Publication Code 636, *Potentials of Aluminium Pressure Die Casting. Application of this technology close to the limits, Magsimal-59, AlMg₅Si₂Mn,* Casting Plant and Technology International, No. 2/2003.

[20] Aluminium Rheinfelden GmbH, Publication Code 545, *Magsimal-59, AlMg₅Si₂Mn, Anwendungsmerkblatt,* February 2004.

[21] Aluminium Rheinfelden GmbH, Publication Code 623, *Qualitätsorientierte Schmelzprüfung in der Aluminiumgiesserei,* Sonderdruck aus Giesserei Heft 23/1987, pages 695–700, 1987.

7
Magnesium alloys for lightweight powertrains and automotive bodies

B. R. POWELL, and A. A. LUO, General Motors
Global Research and Development Center, USA and
P. E. KRAJEWSKI, General Motors Global Vehicle
Engineering, USA

Abstract: This chapter introduces magnesium, the lightest of the structural
automotive metals. It provides an overview of alloy nomenclature and
properties, and the major casting, sheet forming, and extrusion processes.
Descriptions of automotive magnesium applications produced by each
process are provided and there is a summary that describes the challenges in
alloy and process development that need to be overcome if the magnesium
content in automotive sub-system applications is to be increased.

Key words: magnesium, casting, sheet forming, extrusion, hydroforming,
automotive.

Note: This chapter was first published as Chapter 4 'Magnesium alloys
for lightweight powertrains and automotive structures' by B. R. Powell, P.
E. Krajewski and A. A. Luo in *Materials, design and manufacturing for
lightweight vehicles*, ed. P. K. Mallick, Woodhead Publishing Limited, 2009,
ISBN: 978 1 84569 463 0. It is reproduced without revision.

7.1 Introduction

This chapter reviews material characteristics, specific alloys and applications
for magnesium (Mg), the lightest structural metal. The elemental density of
magnesium is 1.74 g/cc, which is one-third less than that of aluminum (Al)
and less than one-quarter that of iron and steel, the other major structural
metals in automotive use today. Magnesium is less dense than most glass
fiber-reinforced automotive polymers and similar in density to that of carbon
fiber composites, although magnesium alloys can cost considerably less.

7.1.1 Magnesium extraction and consumption

Magnesium is a greyish-white metal that makes up 2.7% of the earth's crust
(Okamoto, 1988, pp. 1–3). Due to its high chemical activity, it is never
found as a pure metal in nature. Instead, it occurs commonly as magnesite
($MgCO_3$), dolomite ($MgCO_3.CaCO_3$), brucite ($Mg(OH)_2$), and as the silicates

150

serpentine$((Mg,Fe)_3Si_2O_5(OH)_4)$ and olivine $((Mg, Fe)_2SiO_4)$ (Pidgeon *et al.*, 1946, pp. 4–22). It is also found in sea water, this being a major commercial source of magnesium. The first reported isolation of magnesium metal was accomplished in 1828 by Antoine-Alexander Bussy, who reduced magnesium chloride with potassium (Bussy, 1831). Today, there are three common commercial processes for extracting magnesium: the Dow Process, which extracts magnesium from sea water by electrolysis of magnesium chloride (Emley, 1966, pp. 41–44); the Magnola Process which also uses electrolysis of magnesium chloride, but obtains the chloride through a conversion of magnesium silicates in asbestos tailings (Habashi, 2006, p. 37); and two thermal processes, the Pidgeon Process and the Magnetherm Process (Habashi, 2006, pp 34–36), that use magnesite and/or dolomite ore, which are reduced with ferro-silicon in a retort furnace. For further information about these processes, see Polmear (1999, pp. 3–5).

Magnesium production capacity in 2008 was slightly greater than 700 000 metric tons; about two-thirds of it produced in China, nearly all of that using the Pidgeon process. This tonnage represents a significant increase in production compared with only ten years ago when only 350 000 metric tons were produced. The history of magnesium production is shown in Fig. 7.1. Magnesium has seen a clear upward trend in production over the past 70 years with most of the increase coming in the past ten years.

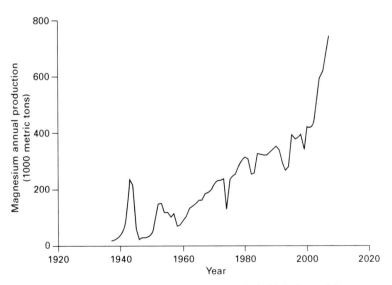

7.1 Magnesium annual production data 1938–2008 (adapted from DiFrancesco and Kramer, 2008).

7.1.2   Overview of magnesium alloys, properties, and processes

For all structural applications, magnesium is alloyed with other metals to provide the proper strength, corrosion resistance, formability, etc. A brief summary (Polmear, 1999, pp 14–15) of some of the typical alloying and impurity elements and their respective effects on magnesium is presented below (letters in parentheses indicate alloy designation in commercial practice):

- Aluminum (A) is the most common alloying element. It improves strength, hardness and corrosion resistance, but reduces ductility. An aluminum content of about 5–6 % yields the optimum combination of strength and ductility for structural applications. Increasing aluminum content widens the freezing range and makes the alloy easier to cast, but more difficult to extrude due to increased hardness.
- Zinc (Z) is used to increase strength. However, it reduces ductility, increases hot-shortness of Mg–Al based alloys, can lead to microporosity and can accelerate corrosion
- Manganese (M) slightly increases yield strength but has no effect on tensile strength. Its most important function is to improve the corrosion resistance of Mg–Al based alloys by removing iron and other heavy-metal elements into relatively harmless intermetallic compounds, some of which separate out during melting.
- Rare earth additions (E) provide precipitation strengthening and can improve both strength and ductility through alteration of crystallographic texture. Rare earths can also improve creep resistance.
- Silicon (S) increases fluidity and slightly improves creep strength.
- Zirconium (K) is an important grain refiner for sand and metal mold, gravity and low pressure casting of magnesium alloys.
- Tin (T) increases ductility
- Lithium (L) reduces density and can improve ductility by creating a cubic structure at over 11% additions. Adding lithium significantly increases cost and reduces corrosion resistance.
- Thorium (H) and yttrium (W) increase creep resistance. The alkaline earths calcium (X) and strontium (J) have also been shown to increase creep resistance.
- Iron (no designation) reduces the corrosion resistance of magnesium alloys and is kept below 50 ppm in all alloys used today (ASTM International, 2008). Similarly, copper (C) and nickel (no designation) reduce the corrosion resistance of Mg alloys. Copper has a letter designation because of its use in ZC63, an early sand casting alloy.

Magnesium alloys are identified using a combination of letters and numbers which describe the major alloying elements and the percentage of those elements. ASTM Standard B 951-08 defines the protocol for use of letters and numbers for naming magnesium alloys (ASTM International, 2008). Per this standard, magnesium alloy families are represented by two letters representing the major alloying elements. (Note: some developmental alloys are represented by three letters or use other nomenclatures, but those included in the ASTM standard are limited to two letters). The letters are shown in Table 7.1. The letters of the alloy represent the two highest concentration alloying elements and are arranged in order of decreasing composition. Following these letters are numbers which show the amount of each of the two alloying elements. The most well-known magnesium alloy is AZ91. This magnesium alloy contains 9 percent (by weight) aluminum and 1 percent zinc. It also contains manganese, but this is not indicated by the name. Other examples include AZ31 which is 3 percent aluminum and 1 percent zinc, and ZK30 which is 3 percent zinc and less than 1 percent zirconium. Following the numbers, a letter may appear which indicates the order of the alloy within the chronology of that particular alloy development. Hence, AZ91B is a more recent version of AZ91A, and so on. These versions may vary in the range and amounts of secondary alloying elements, to satisfy cost targets or to provide some subset of properties or processing benefits. For example, the atmospheric corrosion behavior of magnesium alloys is determined to a great extent by their 'purity' (Hillis, 1983). Thus, the main difference between AZ91C and AZ91D is not in the addition of alloying elements, but instead it is the imposition of maximum values for the impurities copper, iron, and nickel (see Table 7.2).

AZ91E is also shown in Table 7.2. It is the current composition specification for sand cast AZ91, whereas AZ91D is the current composition specification for high pressure die cast AZ91. The lower allowable amount of copper in AZ91E is specified because at the lower rate of solidification during sand casting; the size of copper intermetallics increases the susceptibility of this alloy to micro-galvanic corrosion. Similarly, the higher range of manganese is to react with iron in the melt and remove it from the alloy. Actually the specification for this alloy states that, if either the maximum iron or minimum manganese specification is not met, then there is an alternative specification, setting an upper limit to the iron to manganese ratio.

Table 7.1 Letters representing alloying elements for magnesium

A – aluminum	J – strontium	R – chromium
B – bismuth	K – zirconium	S – silicon
C – copper	L – lithium	T – tin
D – cadmium	M – manganese	V – gadolinium
E – rare earths	N – nickel	W – yttrium
F – iron	P – lead	Y – antimony
H – thorium	Q – silver	Z – zinc

Adapted from ASTM International, 2008.

Table 7.2 Chemical compositions of AZ91C and AZ91D alloys

Element	AZ91C	AZ91D	AZ91E
Aluminum	8.3–9.2	8.5–9.5	8.3–9.2
Copper	0.08	0.025	0.015
Iron	not specified	0.004	0.005
Manganese	0.15–0.35	0.17–0.40	0.17–0.5
Nickel	0.010	0.001	0.0010
Silicon	0.20	0.08	0.20
Zinc	0.45–0.9	0.45–0.9	0.45–0.9
Other elements	0.30 total	0.01 each	0.01 each; 0.30 total

Adapted from ASTM International, 2008.

Magnesium has a hexagonal close packed (HCP) crystal structure, which is different from most structural metals such as aluminum or iron, which are cubic. The hexagonal nature of the magnesium structure makes deformation at room temperature difficult, because there are fewer slip systems for deformation compared with aluminum or iron. Other well known elemental materials which have limited ductility due to their hexagonal structure are titanium and zirconium. Ice also has this property. The mechanical properties of a variety of magnesium alloys are shown in Table 7.3. The individual materials will be discussed in more detail later in this chapter.

7.1.3   Automotive applications of magnesium

Magnesium has been used in a wide range of non-automotive commercial products. The first commercial uses were pyrotechnics, both civilian and military. Due to its high chemical reactivity, magnesium has also had a role in organic chemistry and pharmaceuticals, as well as in the electrochemical industry. Like zinc, magnesium has been used as a sacrificial anode to protect other metals in corrosive environments. Finally, in the metallurgy industry, the major non-component uses of magnesium are as an alloying element for aluminum, to which it imparts strength and corrosion resistance, in steel melt processing to accomplish desulfurization, and in iron melt processing to produce nodular iron. The worldwide demand (in metric tons) for these products in 2007 is shown in Fig. 7.2. Some of the uses of magnesium as a structural material in non-automotive applications are presented in Table 7.4.

The advantages of magnesium as a structural material are summarized in Table 7.5. Some of the key advantages include specific strength, specific stiffness, fluidity, hot formability, machining, and damping. Each of these advantages is briefly addressed in Table 7.5, but it is these advantages that drive consideration for magnesium in automotive applications.

Magnesium has a long history of automotive use dating back to the mid 1930s when it was introduced as an engine block in the Volkswagen Beetle.

Table 7.3 Nominal compositions and typical room-temperature mechanical properties of magnesium alloys

Alloy	Nominal composition, weight percent						UTS MPa	Yield strength		Bearing MPa	Elongation %	Shear strength MPa	Hardness HRB
	Al	Mn	Th	Zn	Zr	Other		Tensile MPa	Compressive MPa				
Sand casting and permanent mold casting alloys													
AZ81A-T4	7.6	0.13	–	0.7	–	–	275	83	83	305	15	125	55
AZ91C-T4	8.7	0.13	–	0.7	–	–	275	145	145	360	6	145	66
EQ21A-T6	–	–	–	–	0.7	1.5 Ag 2.1 Dy	235	195	195	–	2	–	65–85
EZ33A-T5	–	–	–	2.7	0.6	3.3 RE	160	110	110	275	2	145	50
HK31A-T6	–	–	3.0	–	0.7	–	220	105	105	275	8	145	55
WE54A-T6	–	–	–	–	0.7	5.2 Y 3.0 RE	250	172	172	–	2	–	75–95
ZC63A-T6	–	0.50	–	6.0	–	2.7 Cu	210	125	–	–	4	–	55–65
ZK61A-T6	–	–	–	6.0	0.7	–	310	195	195	–	10	180	70
High pressure die casting alloys													
AM60	6.0	0.13	–	–	–	–	205	115	115	–	6	–	–
AS21	1.7	0.4	–	–	–	1,1 Si	240	130	130	–	9	–	–
AZ91D	9.0	0.13	–	0.7	–	–	230	150	165	–	3	140	63
Extruded bars and shapes													
AZ31B-F	3.0	–	–	1.0	–	–	260	200	97	230	15	130	49
AZ61A-F	6.5	–	–	1.0	–	–	310	230	130	285	16	140	60
ZK60A-T5	–	–	–	5.5	–	–	365	305	250	405	11	180	88
Sheet and plate													
AZ31B-H24	3.0	–	–	1.0	–	–	290	220	180	325	15	160	73
HK3A-H24	–	–	3.0	–	0.6	–	255	200	160	285	9	140	68
HM21A-T8	–	0.6	2.0	–	–	–	235	170	130	270	11	125	–

Adapted from ASM International, 1992.

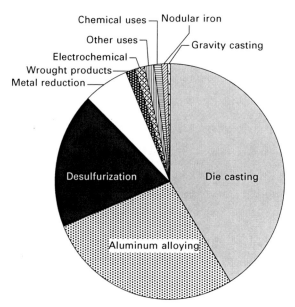

7.2 Magnesium consumption by end use in 2007 (Brown, 2008) (used with permission from TMS).

Table 7.4 Uses of magnesium as a structural material in non-automotive applications

Military and aerospace	Consumer products
• Aircraft air frames • Engines • Transmission cases • Missile skins and frames • Electronic housings	• Power tools • Cameras • Hand luggage • Appliance parts • Cell phones • Portable computers

Since that time, there has been a steady stream of applications covering a wide range of powertrain, chassis, and body structure applications. A pictorial summary of many of these historic and current applications is provided in Fig. 7.3. Most major vehicle components have been made with magnesium either as prototype or production applications. The majority of these applications have been cast components which take advantage of the excellent fluidity of magnesium and the ability to cast very complex and thin walled shapes.

Despite the large number of applications which have been explored or used with magnesium, it remains a relatively small percentage of the materials in a typical vehicle as shown in Fig. 7.4. There are many reasons for the lack of use including the high cost of magnesium, low formability limiting its use in sheet metal components, low corrosion resistance, and limited overall design and manufacturing experience with the material compared to steel.

Table 7.5 Key advantages for improved properties, design, and manufacturing with magnesium

Property	Advantage
• Specific strength	• Magnesium has a specific strength that is similar to cast iron, and similar or greater than many traditional automotive aluminum and thus can provide more mass reduction relative to aluminum.
• Specific stiffness	• Magnesium has a higher specific stiffness than many polymeric materials and composites, thus allowing improved mass reduction.
• Fluidity	• The relatively high fluidity of magnesium allows extremely thin walled castings (1.5 mm), which enhances mass reduction opportunities.
• Hot formability	• Wrought magnesium can be formed into very complex shapes using elevated temperature forming processes.
• Machining	• Machining tools last longer with magnesium than aluminum, reducing costs. The only issue is added care required with machining chips.
• Damping	Magnesium alloys have excellent damping capability compared to other materials making them attractive
• Low temperature properties	• Magnesium does not exhibit a brittle to ductile transition so it can be used at very low service temperatures

Adapted from ASM International, 1999.

Despite these limitations, magnesium has shown significant recent growth in applications as evidenced by the number of applications on the bottom half of Fig. 7.3. The growth and potential for further growth will be discussed in the remaining sections of this chapter.

7.2   Cast magnesium

Of the 370 metric tons of magnesium consumed in 2002, 132 metric tons were as castings; and of these, die castings accounted for 130 metric tons (Webb, 2003). In the automotive industry, high pressure die casting has been the preferred high volume manufacturing method for magnesium. Since 2000, however, other casting processes have been introduced for automotive applications. The growth in these areas is small but increasing. Accordingly, this section will introduce the reader to magnesium casting alloys and casting processes, review the historical and current automotive applications of cast magnesium, and will conclude with a discussion of the challenges and opportunities for automotive magnesium castings.

7.2.1 Cast magnesium alloy nomenclature and alloy families

The protocol for naming magnesium alloys was introduced in the beginning of this chapter. The AZ alloy family (magnesium-aluminum-zinc) is the

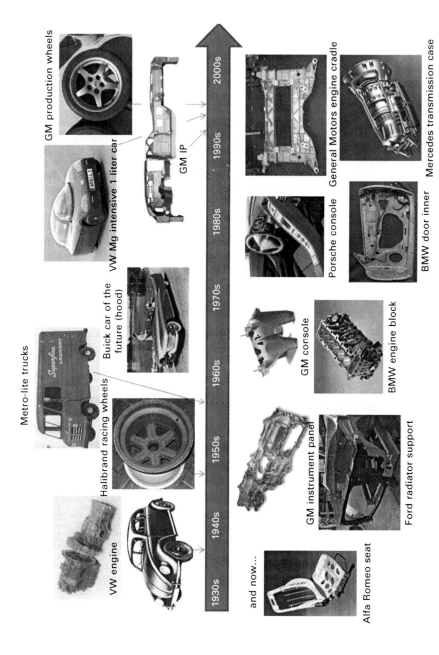

Metro-lite trucks

GM production wheels

VW Mg intensive 1 liter car

GM IP

Buick car of the future (hood)

Halibrand racing wheels

VW engine

1930s 1940s 1950s 1960s 1970s 1980s 1990s 2000s

and now...

Alfa Romeo seat

GM instrument panel

Ford radiator support

GM console

BMW engine block

Porsche console

BMW door inner

General Motors engine cradle

Mercedes transmission case

7.3 Pictorial summary of past and current magnesium automotive applications.

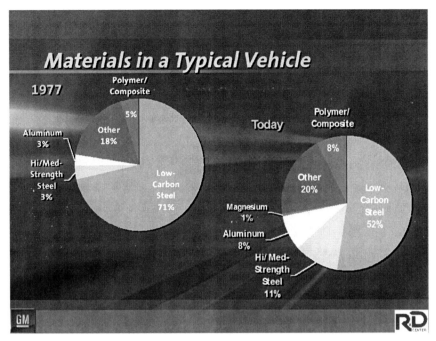

7.4 Breakdown of materials in a typical automobile (after Taub *et al.*, 2007) (used with permission of General Motors).

most well known of the magnesium alloy families because of its wide use in both gravity and low pressure casting, and in high pressure die casting. Some of the alloys in this family have also been used for automotive applications. The AZ alloys have a long and rich history of use in transportation. AZ and the other major families of magnesium casting alloys are shown in Table 7.6. Their application, strengths, and examples of automotive applications are included in the table. The typical casting process is also shown. Specific alloys and nominal compositions for many of the casting alloys are also provided in Table 7.3.

Both AZ and AM alloy families are versatile. They can be cast using a wide range of processes, from gravity and low pressure sand casting to high pressure die casting. As will be seen, they also form the basis of important wrought magnesium alloy families. Among the casting alloys, aluminum is added to magnesium to increase fluidity for castability and to also provide strength to the cast product. Increasing the aluminum content increases tensile strength, but at the expense of ductility. Accordingly, AZ91 alloys are used for applications where strength is the design criterion. At lower aluminum levels, 5 to 6 percent, the ductility of the alloys (AM family) makes them especially attractive for applications requiring crashworthiness; e.g. steering wheels, instrument panels, and seat frames. However, these alloys do not perform well

Table 7.6 Magnesium casting alloy families

Alloy family	Application	Casting process
AZ (Mg-Al-Zn-Mn)	• High strength, low ductility components in thermal environments below 125°C • Alloys – AZ91D and AZ91E • Brake brackets, clutch brackets, covers, transfer cases and gearbox housings, intake manifolds, valve covers, manual transmission cases, wheels, CVT transmission cases, trim, air bag housings	• Gravity or low pressure • Sand or metal mold • High pressure die casting • Squeeze casting • Thixomolding
AM (Mg-Al-Mn)	Lower strength, but higher ductility component; e.g., crash and impact risk; also in thermal environments below 125°C • Alloys – AM50 and AM60 • Steering wheels, seat frames, instrument panels, cross-car beams, trim, roof frame	• Gravity or low pressure • Sand or metal mold • High pressure die casting • Squeeze casting • Thixomolding
AS (Mg-Al-Si-Mn)	Replaces AZ alloys for thermal environments about 125°C due to improved creep strength • Alloys – AS21, AS31, and AS41 • Volkswagen air-cooled engine of 1970's (AS21 and AS41) and currently the Mercedes automatic transmission case (AS31)	• High pressure die casting
AE (Mg-Al-E-Mn)	Higher creep-strength in powertrain operating environment to 150°C, but expensive due to E content • Alloys – AE 42, AE44 • Corvette engine cradle (AE44)	• High pressure die casting
AX and AJ (Mg-Al-Sr/Ca-Mn)	High creep-strength for powertrain components, but potentially at lower cost than E-containing alloys • Alloys – AJ52, AJ62, AXJ530, MRI 153M and MRI 230D (Note: The AXJ and MRI alloys are developmental and do not conform to ASTM standard nomenclature.) • BMW composite engine (AJ62)	• High pressure die casting
Zr refined (Mg-Zn-Zr)	These alloys are all generally high strength and creep resistant, but higher cost, and thus used in aerospace and military applications. The major alloy systems are listed below. For specific examples see Table 7.3 • ZK – (Mg-Zn-Zr) • ZE – (Mg-Zn-E-Zr) • WE – (Mg-Y-E-Zr) • QE – (Mg-Ag-E-Zr)	• Gravity and low pressure casting in sand or metal molds

Adapted from ASM International, 1999.

at elevated temperatures (>125 °C) due to their poor creep resistance. While aluminum improves the castability of the melt, it forms a very low melting point phase with magnesium, ($Mg_{17}Al_{12}$) upon solidification. This phase melts at 450 °C and forms a eutectic with magnesium at 437 °C. Its presence in the microstructure lowers creep resistance of the AZ and AM alloys. While lower Al contents reduce the amount of $Mg_{17}Al_{12}$ formed upon solidification, lower Al is not sufficient to achieve good creep resistance in magnesium alloys. An example is the air-cooled Volkswagen magnesium engine, which used AZ91. It was introduced in the 1970s when engine temperatures were low enough that creep was not a problem. However, as engine performance (and temperature) increased, a new family of alloys was developed with lower aluminum content but containing silicon instead of zinc. This AS family of alloys comprised first AS41 and later AS21 (Mg with 2 percent Al and 1 percent Si), which had better high-temperature properties than AS41. Both alloys relied on the reduced Al content (reduced presence of low-melting $Mg_{17}Al_{12}$) and the formation of Mg_2Si to impart creep strength to the alloy (Hollrigl-Rosta et al., 1980). Although these alloys achieved greater creep resistance, they were not resistant enough for future engine applications.

The creep-resistant alloy AE42 alloy was developed in the 1970s. Rare earth had been shown to impart creep resistance in non-Al containing magnesium alloys because the rare earth formed Mg_9E precipitates in the grain boundaries (Nelson, 1970; Wei and Dunlop, 1992, p. 335). In aluminum-containing alloys, it was discovered that the aluminum reacted with the rare earth and formed $Al_{11}E_3$ precipitates under high pressure die casting conditions and greatly improved creep resistance. However, the rare earth additions increased the cost of the alloy and made it susceptible to hot-cracking (Mercer, 1990). Furthermore, the compressive creep resistance of AE42 decreased abruptly above 150 °C (Sieracki et al., 1996).

While the exact mechanism for the abrupt decrease of creep strength is still debated, Powell and co-workers (2002) reported that this breakdown was accompanied by a decomposition of $Al_{11}E_3$ and the formation of Al_2E and the undesirable $Mg_{17}Al_{12}$ phase. As stated earlier, $Mg_{17}Al_{12}$ is the low-melting-temperature phase that is present in AZ91D, AM60, and AM50 and to which is attributed the poor creep behavior of these alloys. New alloys were subsequently developed that substituted the lower cost alkaline earth elements Ca, Sr, and Ba for rare earth addition, since these elements also formed the $Al_{11}E_3$-type phase, maintained stability above 150 °C, and thus retained creep strength at elevated temperatures suitable for powertrain and underhood applications (Buschow and van Vucht, 1967). This development was subsequently extended to the development of the AXJ series of alloys for creep resistance, specifically AXJ530 (5% Al, 3% Ca, and 0.2% Sr), which demonstrated creep resistance approaching that of aluminum 380 (Powell et al., 2001; Luo et al., 2001). AXJ530 is a developing alloy.

When it is entered into the ASTM tables, its designation will conform to the standard nomenclature. The same can be said for the MRI alloys (see below). Microstructural analysis demonstrated that the Ca addition favored the formation of a grain boundary eutectic structure comprising $(Mg,Al)_2Ca$ instead of $Mg_{17}Al_{12}$ between the primary magnesium grains, and which was more resistant to decomposition during elevated temperature exposure, Fig. 7.5.

Other calcium containing alloys, MRI153M and MRI230D were also developed and commercialized (Aghion *et al.*, 2007). A strontium-containing alloy, AJ62, was also developed (Kunst *et al.*, 2006) and is being used in the BMW composite engine (see Table 7.6).

The remaining alloy families in Table 7.6 contain zirconium. They do not contain aluminum. If present, aluminum will react with the zirconium and precipitate it from the melt (Emley, 1966, pp. 183–185). The zirconium is added during casting of these alloys for grain refinement. None of these alloys are used for high pressure die casting since grain refinement is necessary only at the lower solidification rates of gravity and low pressure casting. The zirconium-refined alloy families have found extensive use in

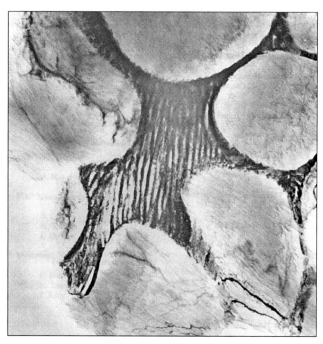

7.5 Analytical electron micrograph of AXJ530 alloy showing primary Mg grains and a grain boundary eutectic structure comprising $(Mg,Al)_2Ca$ and Mg (Powell *et al.*, 2001) (used with permission of SAE International).

aerospace and military applications, but not in the automotive industry due to their high cost (due to the cost of the alloying elements yttrium, silver, and rare earths, and the cost of the zirconium grain refiner). The use of rare earths in high pressure die casting alloys also increases cost, but not as much because these alloys use rare earths in the ratio that they occur in nature, in mischmetal. A typical mischmetal contains 50% cerium, 30% lanthanum, 15% neodymium, and 5% praseodymium. Refinement of mischmetal into individual rare earth elements substantially increases the cost of the alloy. Because the zirconium-refined alloys are not used for automotive applications they will not be discussed further in this section. However, the interested reader is directed to the literature (e.g. ASM International, 1999).

7.2.2   Magnesium casting principles

The steps in casting molten metal consist of preparing the mold or die, melting the metal, melt processing (to remove sources of defects and to modify, refine, or otherwise adjust the chemistry of the alloy), pouring or forcing the melt into the mold or die, solidification and removal of the casting. Casting is a very complex process and even the briefest introduction to its theory and practice is beyond the scope of this book. For further information about casting processes, see *The Metals Handbook Ninth Edition, Volume 15, Casting* (ASM International, 1988). The principles and practice of magnesium casting are generally the same as those for aluminum, but with a few significant differences which will become apparent in this section. The casting characteristics of magnesium are:

- The melting point of pure magnesium is 650 °C. When alloyed, magnesium forms eutectic type systems, (see Fig. 7.6). The melting ranges for die cast magnesium alloys are from 450 °C to 600 °C; those of the zirconium sand casting alloys are typically 550 °C to 650 °C.
- Most magnesium casting alloy families are quite fluid at the casting temperature and are susceptible to oxidation. Due to the small density difference between the metals and their oxides (or dross), any oxide particles or films formed on the melt surface are of neutral buoyancy and thus are likely to become entrapped in the casting. The fluidity of the metal increases the ease of entrapment. These oxides can have a deleterious effect on the mechanical properties of the casting and should be avoided. Magnesium melts are often fluxed and/or passed through filters before entering the casting cavity in the mold.
- An additional method for reducing the occurrence of oxides in the casting involves minimizing the turbulence of the molten metal front during mold filling. This is accomplished in gravity casting (sand or metal mold) by means of unpressurized gating systems. The design of an unpressurized

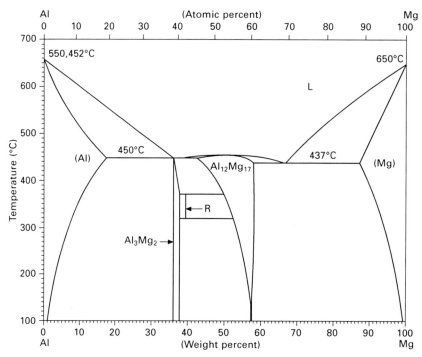

7.6 The magnesium–aluminum phase diagram (after Okamoto, 1998, p.598, with acknowledgement).

gating system is such that the metal front decelerates as it flows through the gating system and into the mold cavity. The deceleration is achieved by continuously enlarging the area of the metal front as it moves though the runners (increasing the runner cross-sectional area) until the runners open into the casting cavity, these openings being termed 'gates.'

- Consistent with controlling metal flow to reduce turbulence is the principle of moving the metal always uphill, at least as much as possible. The principle is difficult to implement in gravity casting and so leads naturally to the idea of low pressure casting, in which the metal is pressure-driven upwards into the casting cavity.

The casting practices described above also apply to aluminum. Magnesium casting methods differ from those for aluminum in two ways. The first is due to the fact that the oxide film that forms on molten magnesium is neither as strong, nor as impermeable, as the film that forms on molten aluminum. This has two effects. The first is an advantage for magnesium because the weakness of the oxide film reduces its detrimental effect on the fluidity of the metal. For this and other reasons, magnesium has a greater fluidity than aluminum, with the result that it is possible to cast more complicated shapes and thinner walls with magnesium than is possible for aluminum; less than

1 mm thick versus greater than 3 mm, respectively, for high pressure die casting. The thin walls that can be cast with magnesium enable greater design flexibility in final components.

The disadvantage of the weak, permeable oxide film is that it creates a safety concern that must be addressed during all steps of the casting process. The film is not an effective barrier to reaction between molten magnesium and other materials in the casting environment, such as the sand mold or the foundry tools and crucibles. Nor is the melt protected from reaction with the atmosphere. These reactions must be prevented since, unlike molten aluminum, which forms an impermeable oxide film and has a low vapour pressure at casting temperatures, e.g. 10E-6 Pa at 660°C. (Magnesium has a very high vapor pressure at its melting point, 360 Pa (Margrave, 1967), and this vapor burns.)

To prevent molten magnesium fires it is necessary to reduce the effective magnesium vapor pressure over the molten metal. One approach is to add beryllium (usually this is already present in the ingot, but it can be added directly to the melt), which, when present even at the 5 ppm level segregates to the melt surface and reacts with the oxide film to strengthen it, thereby reducing its permeability. However, this practice applies only to high pressure die casting because beryllium can interfere with grain refinement, which is necessary for gravity and low pressure casting (Kaye and Street, 1982, p. 146).

Second, fluxes or a reactive cover gas are used during melting and casting. Originally the flux was added as a mixture of sulphur and salts of chloride and fluoride and boron as boric acid. The composition of these flux mixtures was developed for optimum melting, spreading, and viscosity, so that the molten flux would be effective and would remain on the melt surface. Unfortunately, the chloride residues contribute to more corrosion of the casting. Since the 1970s, most fluxes have been replaced with 'cover gas' mixtures containing sulphur hexafluoride (SF_6) at about 0.5% in a carrier gas of argon, air, carbon dioxide, etc. This method provides very good melt protection and became the standard protection method. Unfortunately, SF_6 is now recognized as a powerful greenhouse gas. It is gradually being replaced with sulphur dioxide and various hydrocarbon formulations.

Salt-based flux mixtures are still used to protect the sand mold from reaction with the melt. Magnesium will reduce silicon dioxide and any moisture that is present. Safe practice in the foundry requires the use of these 'inhibitors' in the sand and purging of the sand mold cavity with the cover gas mixture before pouring. Salt fluxes are also kept nearby to spread over magnesium fires that can still occasionally occur.

All foundry tools are kept dry and clean because molten magnesium will react with either the moisture or metal oxides. In the presence of iron oxide, for example, magnesium will reduce the oxide with substantial heat evolution, e.g. thermite. A final word about magnesium fires is in order. Because of magnesium's ability to reduce water and liberate hydrogen, magnesium fires are never fought with water. Instead, flux mixtures are used.

With respect to melt treatment, both magnesium and aluminum reduce and absorb hydrogen from water vapour. When aluminum solidifies, the solubility of hydrogen in it decreases by about 70% (Jorstad, 1993) and results in gas porosity, which usually renders the casting as scrap. The decrease in the hydrogen solubility in magnesium is not so much; less than 20%. Furthermore, the absolute solubility of hydrogen in solid magnesium is much greater than that in aluminum (see Fig. 7.7). The result is that gas porosity in cast magnesium is less of an issue than it is in cast aluminum.

7.2.3   Magnesium casting processes

This section surveys the processes in use today to cast magnesium. Each of the processes has been in industrial use for many years but all of the processes are becoming truly advanced due to several factors: (i) improved process monitoring and control, which is due to advances in sensors and computer software; (ii) more rapid development of the optimum mold/die design resulting from the use of fill and simulation software; and (iii) a greater understanding of the interrelationship between properties, microstructure, and processing, leading to better properties of the castings.

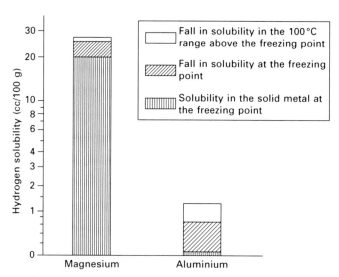

7.7 Solubility of hydrogen in magnesium and aluminum (Kaye and Street, 1982) (with permission from Elsevier).

Gravity and low pressure casting

The distinction between gravity and low pressure casting is that the force of gravity is used to fill the mold in gravity casting. A schematic of a gravity casting mold is shown in Fig. 7.8. Molten metal is poured into a vertical opening (sprue) and flows through the bottom of the sprue and horizontally along a single runner into the casting cavity. The large vertical cavity above the runner in this figure is a riser which provides additional molten metal to 'feed' the casting during solidification. The volume of the metal in the cavity will decrease by several percent due to thermal contraction of the metal and solidification shrinkage. If not continuously fed while the casting is solidifying, porosity will occur in the casting that will degrade the quality of the cast part. As can be seen, molten metal typically flows both up and down in gravity casting despite the need to minimize turbulence.

In low pressure casting, the mold is located above the sprue and metal flows 'up' the sprue and into the runner system and the casting cavity(s), (Fig. 7.9). The metal flow for the arrangement shown in Fig. 7.9 is accomplished by pressurizing the furnace, which is located below the mold. The rate of metal flow is controlled by the rate of pressurization of the furnace. Metal flow can also be directed by electromagnetic pumping, but otherwise the principle of low pressure casting is the same.

High pressure die casting

Metal flow, die filling, and solidification during high pressure die casting are entirely different from gravity or low pressure casting. Molten metal is

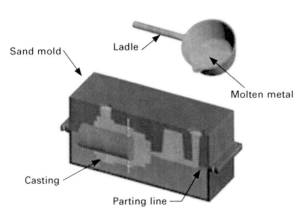

7.8 Schematic of a gravity casting mold showing sprue for receiving poured metal and the runner for horizontal flow of metal from sprue into casting cavity. (with kind permission from Custom Part Inc., www.custompart.net).

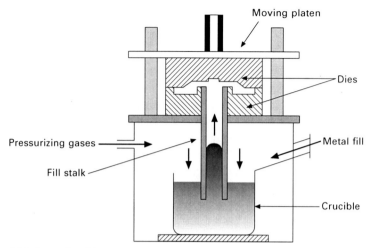

7.9 Schematic of furnace and sprue for low pressure casting. The mold will be located above and in line with the sprue (courtesy Kurtz, Inc.).

injected into the metal mold cavity at very high speed; the linear velocity can be tens of meters per second. Injection is accomplished by pouring the metal into a cylindrical tube (shot sleeve in Fig. 7.10) and using a high speed piston at several m/s to drive the molten metal into the runner system. The result is that the fill time (the time to fill the casting cavity) is measured in tens or hundreds of ms, instead of tens of seconds as in gravity and low pressure casting. These conditions result in the metal being 'virtually' sprayed into the casting cavity where the atomized droplets of metal hit, condense, and solidify rapidly on the internal walls of the casting cavity.

Feeding the casting during high pressure die casting is not accomplished with risers. Rather, extra metal in the shot tube/runners is forced into the full mold under high impact loading; in the order of ~70 MPa. Metal cooling rates are correspondingly much higher; ~100 °C/s during high pressure die casting versus less than 10 °C/s during gravity or low pressure casting. The poor thermal conductivity of sand further lowers the cooling rate to about 1 °C/s in sand casting. The differences in cooling rates drastically affect the solidification rates of the metal and the subsequent tensile properties of the casting. Most of the structural integrity of the high pressure die cast magnesium part is in the skin of the casting, the skin being the thin external surface of the casting which is characterized by very low porosity and very fine grain size. Extensive porosity can occur beneath the skin (core) due to the filling and solidification conditions. As long as the core is not exposed by machining away the skin or other means, the mechanical and sealing integrity of the casting can be maintained. Often however, the design of the part will require some machining; thus controlling the amount and location of porosity in the cast product is critical.

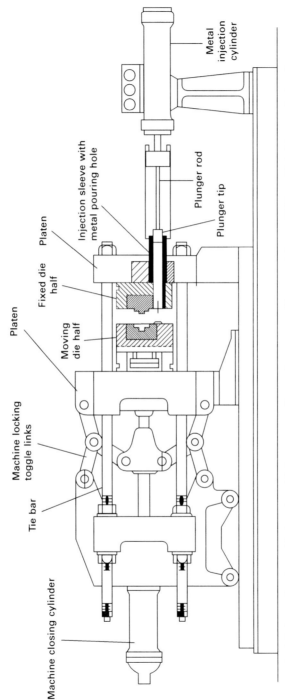

Machine closing cylinder

Tie bar

Machine locking toggle links

Platen

Moving die half

Fixed die half

Platen

Platen

Injection sleeve with metal pouring hole

Plunger rod

Plunger tip

Metal injection cylinder

7.10 Schematic of a cold chamber, high pressure die casting system (Kaye and Street 1982, p. 6) (with permission from Elsevier).

Advantages of high pressure die casting include higher productivity because the filling and solidification rates, as well as the time to eject (remove) the part from the mold, are much faster than for gravity or low pressure casting. The chief disadvantage is gas entrapment in the casting, which occurs when molten metal is injected into the cavity and results in casting defects which lower the strength and ductility of the part. In addition to the casting defects, the presence of entrapped gas also precludes heat treating to improve properties. Vacuum die casting, where the die cavity is evacuated prior to and during metal fill has been developed (Vinarcik, 2003, pp. 29–49; Luo and Sachdev, 2008), but the level of vacuum achieved is still not enough to eliminate resultant porosity. The development of 'supervacuum die casting,' may further reduce porosity and improve properties (Brown *et al.*, 2009). At the time of writing, essentially all of the magnesium parts cast for use in the automotive sector are high pressure die cast. If vacuum and supervacuum die casting can be commercialized, the range of automotive applications for magnesium will increase.

Squeeze casting

Squeeze casting is a hybrid of low pressure casting and high pressure casting, and it has the potential to completely eliminate the gas defects associated with high pressure die casting, and to enable heat treatment of the castings. In squeeze casting, the die is filled slowly with metal to maintain laminar flow. Once the cavity is full, the pressure on the melt is increased to over 100 MPa and maintained to feed the casting to compensate for shrinkage until the casting has solidified. Die design for squeeze casting is different from that for die casting, and includes thick gates and a large shot end biscuit to ensure that the gates do not freeze before the casting, in the cavity has solidified and to ensure feeding the shrink during solidification. Other advantages of squeeze casting are contained in Table 7.7 (Kainer and Benzler, 2003).

Table 7.7 Advantages of squeeze casting for magnesium

• Reduced porosity
• Hot cracking prevention for alloys with wide freezing ranges
• Increased strength and ductility due to:
– Fine-grained microstructure
– Defect-free microstructure
• Possibility of heat treatment to improve properties
• Offers a wider range of alloy choices, including those that may be difficult to cast by other means
– Creep-resistant alloys
– Thixotropic melts
• Metal matrix composites can be produced

Source: Kainer and Benzler, 2003.

The limitations of squeeze casting include less flexibility in part geometry, lower productivity, high machining requirements, and greater cost. However, as a potential casting process for safety-critical parts in automotive systems such as space frame joints, squeeze casting may find its niche.

Semi-solid casting

Semi-solid casting is a casting process that involves filling a mold with the metal in a partially molten state in which globules of solid are homogeneously dispersed in the liquid. The benefits of semi-solid casting are: (i) reduced shrinkage due to the lower casting temperature of the melt, (ii) low gas porosity enabling heat treating of the castings, (iii) high mechanical properties due to the uniquely fine microstructure of the resulting castings, and (iv) extremely fine surface finish. In some cases semi-solid castings can compete with forgings. Semi-solid aluminum casting was developed in the 1970s by Flemings (1974), it was referred to as rheocasting. In that process, an alloy containing two phases, the alpha phase with a higher melting point than the eutectic, is stirred at a preselected temperature between the liquidus (all liquid above this temperature) and solidus (all solid below this temperature) of the starting alloy to provide about 35% liquid, to obtain the homogeneous distribution of the solid phase dispersed in the liquid (Fig. 7.11). The particular distribution and volume fractions of the two phases render the material thixotropic, where its viscosity decreases with shear rate, to enable the mixture to be injected at high rates into a die cavity while maintaining laminar flow. The key to doing this successfully is thermal management of the runner system and the die cavity. The process later evolved into thixocasting, where carefully prepared electromagnetically stirred billets that contain globular alpha aluminum are dispersed within a lower melting point eutectic phase. These billets are reheated to a temperature to melt the eutectic and inject the mostly solid mass at high velocity into the die cavity.

A yet third process was developed under the umbrella of semi-solid forming, called thixomolding, and has gained greater use for magnesium compared to the other two processes, which are used primarily for aluminum. In thixomolding (see Fig. 7.12), magnesium chips are fed into a modified plastic injection molding machine and are melted in the barrel while an internal screw advances the mass towards the shot end (LeBeau *et al.*, 1998). By the time the material mass reaches the shot end, it contains about 5–10% solid, which is injected in to the die cavity. The much lower metal temperature compared to die casting allows the casting of extremely thin wall magnesium castings.

Because of the demands of die thermal management, thixomolding has generally been limited to small parts such as cell phones and other consumer electronics such as computer cases, which have the common characteristic

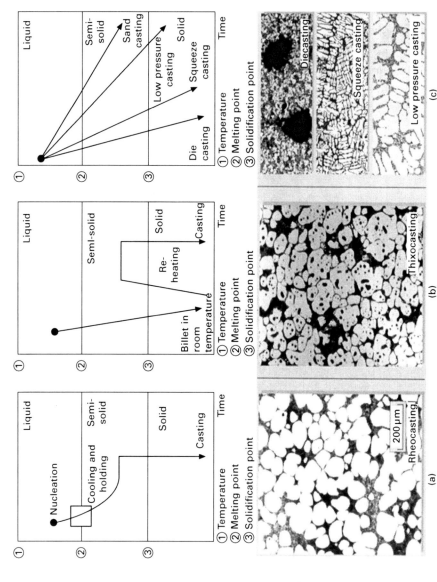

7.11 Comparison of microstructures resulting from various casting processes (courtesy Ube Machinery, Inc.).

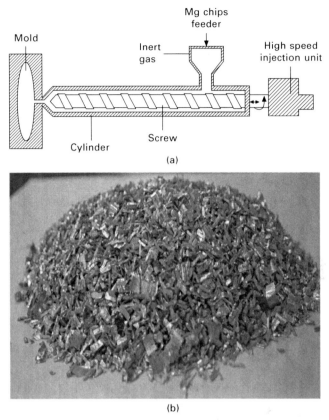

(a)

(b)

7.12 Schematic of (a) the thixomolding equipment with (b) a representative image of the metal chips/feed used in the process (used with permission from R. Decker,Thixomat).

of being very thin wall castings. Also, the molding temperature being lower than other casting processes, the mold life can be extended and the cast parts have fine and clean surfaces which are cosmetically appealing. Finally, the lower fill rate of the semi solid process leads to less entrapped gas and shrinkage, giving better properties and enabling post casting heat treating.

Over time, as die thermal management techniques evolve, it is expected that the advantages of thixomolding will extend to automotive components.

7.2.4   Automotive applications of cast magnesium

One of the earliest applications of magnesium as an automotive material was a sand cast crankcase on the 1931 Chevrolair. Commercially viable applications appeared in London in the 1930s (Emley, 1966, p. 828). These applications included lower crankcases for city buses, and transmission housings for tractors (see Fig. 7.13). The crankcase shown in Fig. 7.13 weighed 26 kg. Over

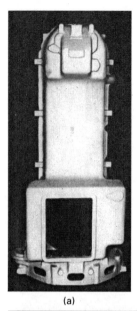

(a)

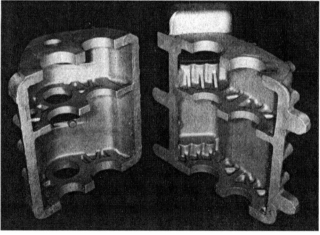

(b)

7.13 Fully commercialized cast magnesium powertrain components used in England in the 1930s, (a) a lower crankcase and (b) a transmission housing (Emley, 1966) (used with permission from Elsevier).

500 000 transmission housings were cast. Both components were sand cast. Other magnesium components included gear boxes, clutch housings, sumps, chassis parts, wheels, and truck body components, usually sand cast with Commercial C alloy, AZ81. Crankcases and housings were also produced in Germany, but by high pressure die casting (HPDC). The alloy used for these castings was AZ81. Together, these components weighed 17 kg, which was

50 kg less than the iron components they replaced (Schumann and Friedrich, 1998). An example of a large HPDC casting is shown in Fig. 7.14.

Magnesium usage grew throughout the 1930s and then grew exponentially to nearly 250 000 metric tons production during World War II as mass reduction, desulfurization, and alloying became critical in both land and air weaponry (see Fig. 7.1). After the War, magnesium demand collapsed and remained low until the mid-1950s. With the introduction of the Volkswagen Beetle, automotive magnesium consumption again accelerated. Worldwide magnesium for all applications reached 240 000 metric tons in 1971, of which 42 000 metric tons were for the Beetle air-cooled engine and gearbox, which together weighed about 20 kg (Friedrich and Schumann, 2001). However, then several factors emerged and combined to cause the reduction and eventual elimination of magnesium as a structural powertrain component. These factors included greater power requirements for the engine, which increased both its operating temperature and load and which ultimately resulted in the conversion of the engine from air cooling to water cooling; the AZ81 alloy, and later the AS41 or the AS21 alloys, lacked sufficient creep resistance in the required operating environment (Hollrigl-Rosta, 1980). The use of water cooling put magnesium at a disadvantage compared with other engine materials because of its poor corrosion resistance. At that time, the effect of iron, copper, and nickel impurities (in ppm amounts) on promoting the corrosion of magnesium had not been recognized (Hillis, 1983). By the time the 'high purity' alloys AZ91D and AM60B, which replaced AZ91C and AM60A, respectively, were developed in the 1980s, the cost of magnesium alloys had begun to increase, and the use of magnesium in automotive

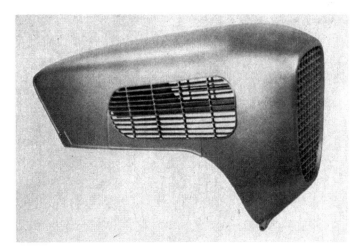

7.14 A high-pressure die cast tractor hood. The nominal wall thickness for this casting was 2.2–2.5 mm. (Emley, 1966) used with permission from Elsevier).

applications decreased although some applications remained. These included cam covers, brackets, and manual transmission cases, which used the new, greater corrosion-resistant magnesium alloys that resulted from the work of Hillis and others. With time, high pressure die cast automotive applications have increased, as shown in Section 7.1. A lot of the increase is attributed to the many advances made in the development of casting fill and solidification software, in addition to the high purity alloys that became available. The simulation tools have enabled a much broader range of cast products, both from the perspective of complexity and properties. These applications, which accounted for a better than ten percent annual growth in magnesium usage per year throughout the 1990s, included instrument panel structures and cross-car beams, seat frames, steering wheels, intake manifolds, and numerous brackets and covers. General Motors usage of magnesium in 2005 is shown in Fig. 7.15.

Other significant applications include the Mercedes 7-speed Tiptronic automatic transmission case (Greiner *et al.*, 2004) (see Fig. 7.16), the BMW composite engine (Hoeschl *et al.*, 2006) (Fig. 7.17), and the Corvette engine cradle (Fig. 7.18) (Li *et al.*, 2005). The BMW composite engine is an inline block, and uses an aluminum inner casting to carry the head bolts, bulkheads, water cooling passages, and cylinder bores. The BMW engine block is 57 percent lighter than the iron block it replaced. The Corvette engine cradle was the result of a project which was jointly sponsored by the US Department of Energy (DOE) and the US Council for Automotive Research (USCAR), and led by the OEMs in the US. Two other USCAR projects that are ongoing are also pushing the envelope for automotive uses of magnesium. The first is the Magnesium Powertrain Cast Components (MPCC) Project which has the objective of demonstrating the readiness of magnesium alloys for completely replacing the major aluminum components of a V block engine

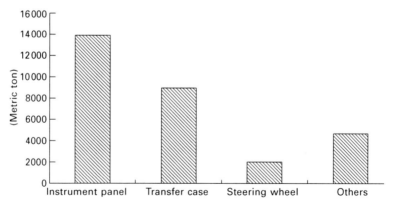

7.15 Usage of magnesium in General Motors North American vehicles in 2005 (courtesy World Wide Purchasing, General Motors Corporation).

7.16 The Mercedes 7-speed automatic transmission case (after Greiner *et al.*, 2004) (used with permission of Daimler).

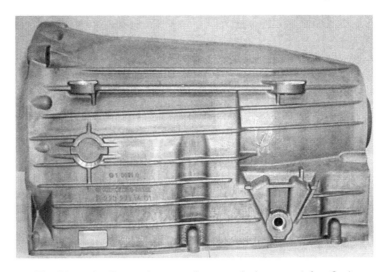

7.17 The Mercedes 7-speed automatic transmission case (after Greiner *et al.*, 2004) (used with permission of Daimler).

(Powell *et al.*, 2005). The MPCC cylinder block achieved a mass reduction of 25% (29% for all of the cast aluminum components which were replaced by magnesium). A prototype engine made with a low pressure sand cast cylinder block, a Thixo-molded rear seal carrier, and high pressure die cast oil pan and front cover, with all other parts carried over from the baseline aluminum engine, has been completed (Powell, 2009) (see Fig. 7.19). The prototype engine is being dynamometer tested for durability. The other USCAR project is the Front End Project (Luo and McCune, 2008) which is

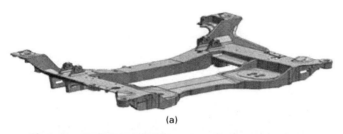

(a)

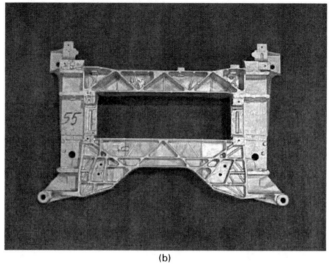

(b)

7.18 The Corvette engine cradle (a) in schematic and (b) actual casting (Courtesy General Motors).

described later. Briefly, that project seeks to create a completely magnesium front end structure, thereby reducing mass significantly (see the schematic of the front end in Fig. 7.20). Projects such as these are good examples of collaborative projects that are advancing the field. For more information about the US DOE Lightweighting Program, see Carpenter *et al.*, 2008.

7.3   Sheet magnesium

7.3.1   Alloy families, nomenclature and properties

The number of commercially available magnesium sheet alloys is very limited. The most commonly used alloy is the aluminum and zinc containing AZ31B. Other alloys which are currently commercially available are AZ61, HM21, HK31 and ZM21. The materials are typically available in either an annealed, O temper or in a partially hard H-24 temper. Typical microstructures of AZ31 in these two temper conditions are shown in Fig. 7.21. The annealed O temper material has a homogeneous distribution of grains with clearly visible

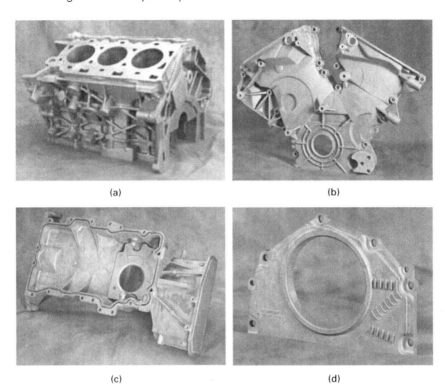

(a)   (b)

(c)   (d)

7.19 Magnesium powertrain components from the USCAR Magnesium Powertrain Cast Components Project; (a) cylinder block, (b) front engine cover, (c) oil pan and (d) rear seal carrier (Powell, 2009) (used with permission from Ford Motor Co.).

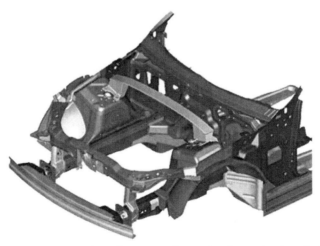

7.20 Schematic of the front end of a production sedan (used with permission from General Motors Inc.).

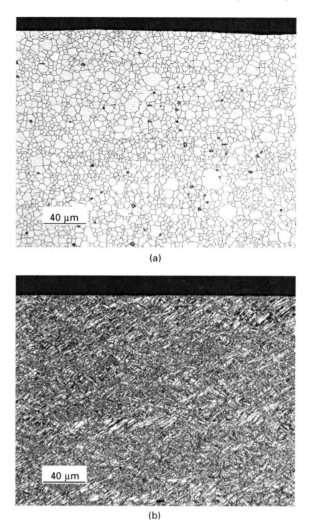

(a)

(b)

7.21 Typical microstructures in AZ31 sheet showing (a) annealed,
O temper and (b) as-rolled, H-24 temper.

boundaries, while the, as rolled, H24 temper material is highly worked and it
is difficult to discern individual grains. In the O temper, the AZ31 material
provides similar yield strength to work hardenable aluminum sheet alloys such
as AA5754 or AA5182, but lower strength than age-hardenable alloys such
as AA6111. One issue with the mechanical properties of sheet magnesium is
that the properties are typically anisotropic, meaning the strength or ductility
varies with direction on the sheet. For example, yield strength is typically
approximately 10% lower transverse to the rolling direction compared with
parallel to the rolling direction. The anisotropy is the result of two main

features, (i) irregular or non homogeneous grain structure and (ii) texture differences, between the rolling and transverse to rolling directions. An example of grain inhomogeneity is shown in Fig. 7.22, which shows regions of very fine grains oriented in bands through the material.

7.3.2   Magnesium sheet forming processes

Traditionally, sheet magnesium has been made by ingot metallurgy and subsequent hot rolling. This can be a very expensive process, as the limited formability of magnesium at room temperature necessitates that the reduction of the ingot to the final sheet be done at elevated temperatures. As a result, the conversion cost (cost to convert raw magnesium material into sheet product) is higher than for converting steel or aluminum into sheet. Recently, a significant amount of work has been ongoing globally to develop a lower cost, strip casting process whereby magnesium sheet can be cast to almost net shape, and then require minimum rolling to the final gage. More information on this process can be found elsewhere (Krajewski *et al.*, 2007). Because of the high cooling rates in strip casting, these materials can have a finer grain size than ingot cast materials which can be an advantage to both strength and formability. An example of a fine grain structure produced by strip castings is shown in Fig. 7.23.

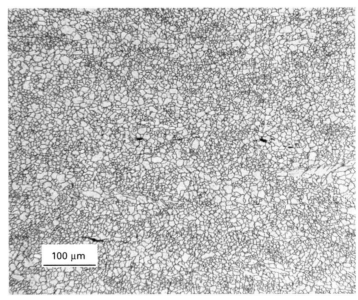

7.22 Micrograph of rolled and annealed AZ31 material showing inhomogeneous grain structure with bands of fine grains.

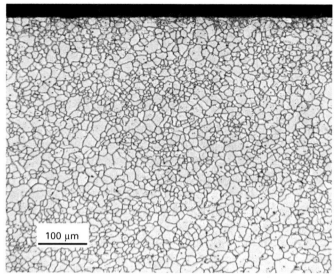

7.23 Micrograph of strip cast AZ31 material very fine grain structure.

7.3.3   Forming processes using magnesium sheet

The majority of the processes used to convert sheet metal into automobile components occur at room temperature including, stamping, flanging, bending, hemming and trimming. The processes are very robust with high formability materials such as mild steel, and have also been used for less formable materials such as aluminum and high strength steel. Unfortunately, the limited formability of magnesium due to its HCP structure (discussed earlier) makes the use of these processes very difficult. An example of the problem is shown in Fig. 7.24 where the results of a forming trial using a simple rectangular pan for a mild steel and AZ31B magnesium sheet are compared. The pan could be formed to a 125 mm depth with the steel, but split after only about 12 mm with the magnesium (Krajewski *et al.* 2006).

Another example of the limited formability of magnesium sheet can be seen by comparing the minimum bend radius for magnesium with those of other materials. *The Metals Handbook* (ASM International, 1998) suggests that for a 90 degree bend, the minimum bend radius for magnesium sheet is between $5t$ and $13t$, where t is the metal thickness. This depends on the alloy and temper chosen, with AZ31B-O being the best. For comparison, aluminum and steel can achieve bending radii approaching $1t$ for a 180 degree bend which is essentially bending the material flat over itself.

The limited formability at room temperature with magnesium sheet necessitates the use of elevated temperature forming processes. Increasing the temperature significantly increases bendability, thereby reducing the achievable bend radius. This is summarized in Table 7.8, taken from *The*

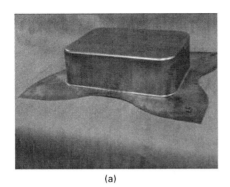   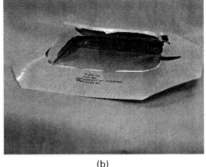

(a)   (b)

7.24 Comparison of a forming trial on a 125 mm deep pan with (a) mild steel and (b) AZ31B magnesium.

Table 7.8 Recommended minimum radii for 90° bends in magnesium sheet

Alloy and temper	Forming temperature (bend radii as multiples of sheet thickness)							
	20 °C (70°F)	95 °C (200°F)	150 °C (300°F)	205 °C (400°F)	260 °C (500°F)	315 °C (600°F)	370 °C (700°F)	425 °C (800°F)
AZ13B-O	5.5tp	5.5t	4t	3t	2t	...	...	...
AZ13B-H24	8t	8t	6t	3t	2t	...	...	...
HK31A-O	6t	6t	6t	5t	4t	3t	2t	1t
HK31A-H24	13t	13t	13t	9t	8t	5t	3t	...
HK31A-T8	9t	9t	9t	9t	9t	8t	6t	4t

Adapted from ASM International, 1998.

Metals Handbook (ASM International, 1998). All of the applications described previously were created using some variant of an elevated temperature forming process. A brief summary of the relevant elevated temperature forming processes now follows.

Superplastic forming

Superplastic forming (SPF) involves slowly forming a sheet of material in a single-sided tool or die using gas pressure at a temperature of about 500 °C. A typical forming cycle for SPF is approximately 30 minutes or longer. The process requires material which has been specially processed to have a fine grained or dual phase microstructure (ASM International, 1998). Since the 1970s, SPF has also been used by the automotive industry to produce complex, lightweight *aluminum* panels for niche products with annual volumes of less than 1000 units (Barnes, 2002). The technology has been optimized around niche vehicles where tooling costs are extremely low, cycle times are long, labor content and material price are high, and parts can be reworked after production because each vehicle is handcrafted. Superplastic formability trials on automobile components have been carried out on a number of magnesium alloys including AZ31B, ZK10, and ZE10 (Barnes, 2007, pp. 440–454). The excellent high temperature ductility of magnesium sheet enables this process to be used, often without special processing of the magnesium sheet. The technology should be able to make viable low to niche volume automotive panels with magnesium sheet.

Quick plastic forming

Quick plastic forming (QPF) was developed as a gas forming technology that could be used in the mainstream automotive industry (Rashid *et al.*, 2001; Schroth, 2004, p. 9; Krajewski and Schroth, 2007, pp. 3–12). As in SPF, forming is done at elevated temperatures using gas pressure. The temperatures used in QPF (~450 °C) are typically lower than those used for SPF (~500 °C). However, much faster forming cycles are achieved (~1–5 minutes). The faster cycle time is achieved through process automation, thermal control of the sheet metal and dies, and a number of other improvements detailed elsewhere (Krajewski and Schroth, 2007, pp. 3–12). These improvements enable the production of components at automotive-type volumes, up to 75 000/year.

As with SPF, QPF has also been implemented to produce *aluminum* components for the Chevrolet Malibu Maxx and aluminum decklids for the Cadillac STS. The QPF process has been used to produce many prototype magnesium inner closure panels using the same bill of process (BOP) which was used for the production of aluminum QPF panels (Carter *et al.*, 2008). As with SPF, this technology is ready for use with magnesium sheet but at much higher volumes. Examples of QPF prototype products are shown in a later section.

Warm forming

Warm forming refers to elevated temperature stamping that gives higher forming rates but lower formability than processes such as QPF or SPF. It is typically carried out at temperatures between 200 °C and 350 °C using heated, matched die sets, and has been the standard process for stamping complex shapes from sheet and plate magnesium in non-automotive industries including aerospace components, luggage, and computer cases. Warm forming guidelines for draw depths, stretching limits, and bending radii, have been summarized by Emley (1966, pp. 584–603). Spinning, hammer forming, and rubber pad forming have also been used at 'warm' temperatures. Warm stamping of magnesium can enable a wide range of moderately complex automobile components to be formed. An example of a part formed with this process is the door inner panel shown in Fig. 7.25, formed using AZ31B (Krajewski *et al.*, 2006). The issue currently preventing warm forming from

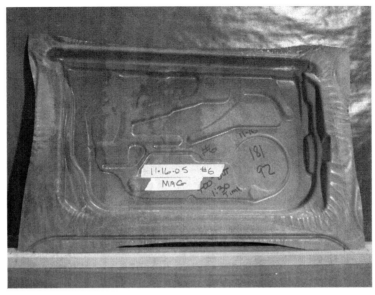

7.25 Magnesium door inner panel made by warm forming (Krajewski *et al.*, 2006).

being used is the lack of an integrated warm forming process. This includes the need to rapidly heat blanks, control die temperature and dimensions, as well a developing a lubricant.

Thermohydroforming

Another process for elevated temperature forming of magnesium sheet is thermohydroforming. Thermohydroforming is a combination of warm forming and traditional sheet hydroformimg (Groche *et al.*, 2002).

Using heated synthetic oil as a forming medium, the magnesium sheet blank is formed into the desired shape using the fluid pressure. There are many varieties and applications of conventional room temperature hydroforming, but thermohydroforming remains only a research topic due to limitations in developing suitable oil which is stable at the warm forming temperatures, and handling the oil in a safe manner in a production environment.

Warm hemming

A critical process for assembling automotive closure panels is hemming, in which an outer panel (exterior of the vehicle) is bent to 180° over an inner panel to create a closure assembly. For a uniform 180° flat hem, the flange die radius should be approximately equal to or less than the thickness of the outer panel, *t*, which would allow for an outside metal radius in the order of 2*t* or less. In steel and aluminum, this process occurs at room temperature. Bending wrought magnesium alloys with such a sharp radius however, would cause unacceptable cracking on the outer surface unless the bending was performed at an elevated temperature. A recent study (Carsley and Kim, 2007, pp. 331–338) showed that successful hemming of AZ31B-O, required temperatures of 270–280 °C to achieve an acceptable condition with good surface appearance. Bending at these temperatures ensures that the material is annealed, that the grain structure is recrystallized, and that additional slip systems are activated to accommodate the uniform bending strain without plastic localization (necking) or cracking failure at the outer surface of the panel. While this laboratory study provided encouragement that warm hemming can be performed, a viable production method is needed.

Warm clinching

In addition to hemming, clinching is a critical technology to assemble automobile components. For example, mechanical clinching has been used extensively to attach reinforcements to aluminum inner closure panels. The extremely high localized deformation in mechanical clinching operations makes them very difficult to execute at room temperature with magnesium. Warm clinching of magnesium sheet to either aluminum sheet or magnesium sheet has been demonstrated on a laboratory scale (Behrens and Hubner, 2005, pp. 59–62). In this work, clinching equipment was modified by the addition of electrical resistance heaters to heat the punch and die to 390 °C. Water cooling and insulation were added to keep the rest of the equipment cool. When clinching magnesium (punch side) to aluminum (die side), a minimum temperature of 275 °C was required to form good joints. For clinching magnesium to magnesium, a minimum of 200 °C was required. As with hemming, robust, production-viable methods of warm clinching need to be developed.

7.3.4   Automotive applications of magnesium sheet

Magnesium sheet has been used as a structural material in the transportation industry since World War II. The most famous application was the B36 bomber, which contained 9000 lbs of magnesium sheet (Barnes, 1992, pp. 29–43). The first commercial ground transportation applications were developed in the early 1950s. Metro-lite trucks were manufactured between 1955 and 1965 and featured magnesium sheet panels as well as structures made from magnesium plate and extrusions (Barnes, 1992, pp. 29–43). These trucks had increased payload capacity and were excellent applications for magnesium because they did not require extreme formability. However, sheet magnesium has not been used in high volume production in the mainstream automobile industry. The only current production application of magnesium sheet is the center console in the low volume Porsche Carerra as shown in Fig. 7.26.

While production applications are limited, numerous prototype components have been made using sheet magnesium. General Motors made prototype hoods for the Buick LeSabre in 1951, various body panels for the Chevrolet Corvette SS Race Car in 1957, and hoods for the Chevrolet Corvette in 1961 (see Fig. 7.27).

More recently, Volkswagen has made a prototype hood for the Lupo, shown in Fig. 7.28 (Moll *et al.*, 2004). GM has made numerous panels including a hood, door inner panel, decklid inner, liftgate, and various reinforcements, some of which are shown in Fig. 7.29 (Krajewski, 2001; Verma and Carter, 2006; Carter *et al.*, 2008). Chrysler LLC has performed a number of studies using magnesium sheet including the inner panel shown in Fig. 7.30 and a magnesium intensive body structure (see also Logan *et al.*, 2006).

The majority of the applications just discussed were 'inner' panels which create the structure of the vehicle closures but are not visible on the outside of the vehicle. This is due to two factors. First, the surface quality of the

7.26 Sheet magnesium center console cover in Porsche Carrera GT automobile (copyrighted by Dr Ing. H.c. F. Porche AG. Used with permission).

(a)

(b)

7.27 (a) 1951 Buick LeSabre concept car with magnesium and aluminum body panels, (b) 1961 Chevrolet Corvette with prototype hood made from magnesium sheet, and (c) 1957 Chevrolet Corvette SS Race Car with 'featherweight magnesium body' (copyright 2007 GM Corp. Used with permission, GM Media Archive).

(c)

7.27 Continued

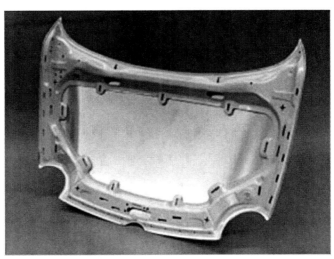

7.28 VW Lupo magnesium hood (Moll *et al.*, 2004) (used with permission from John Wiley and Sons).

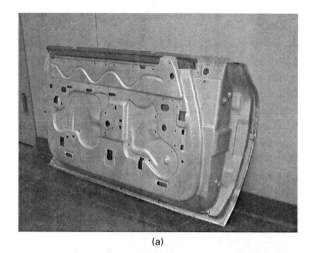

(a)

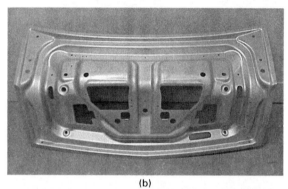

(b)

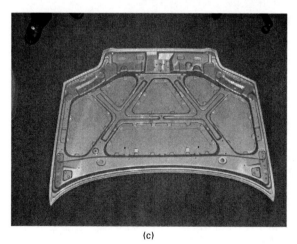

(c)

7.29 Magnesium sheet panels formed recently by General Motors: (a) door inner panel, (b) decklid inner panel, and (c) hood. (Krajewski 2001; Verma and Carter, 2006; Carter *et al.*, 2008).

7.30 Inner panel drawn by Daimler-Chrysler using magnesium sheet (used with permission from Chrysler Corp.).

currently available magnesium sheet requires significant finishing compared to aluminum or steel, and second, the limited formability at room temperature makes assembly processes for outer panels, such as hemming, difficult. This means that the first commercial applications for magnesium sheet are likely to be inner panels.

7.4   Extruded magnesium

7.4.1   Alloys and properties

Table 7.9 lists the nominal composition and typical room-temperature tensile properties of common extruded magnesium alloy tubes (ASM International, 1999; Timminco Corporation, 1998; Luo and Sachdev, 2007, pp. 321–326). Of the commercial extrusion alloys, AZ31 is most widely used in non-automotive applications. The higher aluminum content alloys, AZ61 and AZ80, offer greater strength than the AZ31 alloy, but have much lower extrudability. The high-strength Zr-containing ZK60 was designed for applications in racing cars and bicycles, such as wheels and stems (Timminco Corporation, 1998). The extrusion speed of ZK60A tubes is extremely low, rendering it uneconomical for automotive applications.

AM30 (3 percent Al, 0.4 percent Mn), a new extrusion magnesium alloy developed by General Motors (Luo and Sachdev, 2007, pp. 321–326), is aimed to provide a good balance of strength, ductility, extrudability and corrosion

Table 7.9 Nominal compositions and typical room-temperature tensile properties of extruded magnesium alloy tubes

Alloy	Temper	Composition (wt per cent)				Tensile properties		
		Al	Zn	Mn	Zr	Yield strength (MPa)	Tensile strength (MPa)	Elongation (percent)
AZ31	F	3.0	1.0	0.20	–	165	245	12
AZ61	F	6.5	1.0	0.15	–	165	280	14
AZ80	T5	8.0	0.6	0.30	–	275	380	7
ZK60	F	–	5.5	–	0.45	240	325	13
ZK60	T5	–	5.5	–	0.45	268	330	12
AM30	F	3.0	–	0.40	–	171	232	12

(adapted from ASM International 1999; Luo and Sachdev, 2007; Timminco Corporation, 1998)

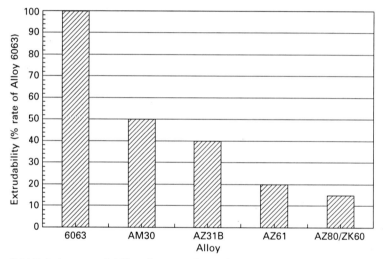

7.31 Relative extrudability of magnesium alloys compared to alloy 6063 (Luo and Sachdev, 2008).

resistance. Depending on the extrusion temperature used, the AM30 alloy was shown to extrude 20–30% faster without visible defects compared to the conventional AZ31 alloy, (Luo and Sachdev, 2007, pp. 321–326). Figure 7.31 summarizes the relative extrudability of magnesium alloys compared to aluminum alloy 6063, based on information from the literature (ASM International, 1999; Timminco Corporation, 1998; Busk, 1987); and the authors' experience (Luo and Sachdev, 2007, pp. 321–326).

7.4.2   Extrusion processes

Special equipment is not required to extrude magnesium alloys. The alloys can be warm or hot extruded in hydraulic presses to form bars, tubes, and a wide variety of profiles (ASM International, 1999). While hollow magnesium extrusions can be made with a mandrel and a drilled or pierced billet, it is generally preferable to use a bridge die where the billet is broken into three to four streams that get their internal shape as the streams flow over the supporting mandrel whose end face sits essentially flush with the die exit face that contains the external profile of the extrusion.

The hydrostatic extrusion process, typically used for copper tubing fabrication, is a much faster extrusion process compared with conventional direct extrusion. It was recently reported that seamless magnesium tubes were extruded using the hydrostatic process at speeds up to 100 m/min, due to the absence of friction between the billet and container since the billet is suspended in hydraulic oil (Savage and King, 2000, pp. 609–614). Although the process is capable of extrusion ratios up to 700, the outer diameter of the tubes produced by this process is limited to about 45 mm, even with a large 4000-ton press (Savage and King, 2000, pp. 609–614).

7.4.3   Magnesium tube bending

Bending at room temperature

The rotary draw bending process, as shown in Fig. 7.32, is generally used for bending aluminum and magnesium tubes. In this process, the tube is clamped against the bend die insert with the clamp die, and a multi-ball steel mandrel is positioned inside the tube. The pressure die holds the collet end of the tube against the wiper die as the bend die rotates and the tube is drawn forward. Bending of conventionally extruded magnesium AZ31 and AM30 tubes at room temperature is generally unsuccessful; a 2D/90° bend could not be consistently achieved (Luo and Sachdev, 2005, pp. 477–482; Luo *et al.*, 2005, pp. 145–148). Figure 7.33 compares the longitudinal cross section micrographs of an AZ31 tube before and after a 30° bend at room temperature. The micrograph in Fig. 7.33(b) clearly shows the onset of strain localization initiating a crack prior to fracture. There is no evidence of any grain elongation, even near the outermost surface where the strain is the largest, suggesting that twinning is the predominant deformation mode at room temperature.

Bending at elevated temperatures

A moderate temperature (100–200 °C) bending process was developed at GM for magnesium tubes, using a Pines rotary draw bending machine

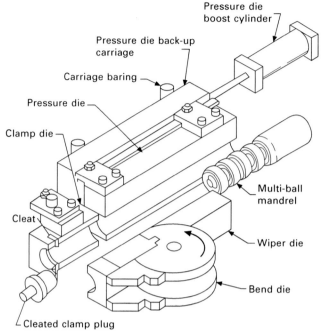

Pressure die
boost cylinder

Pressure die back-up
carriage

Carriage baring

Pressure die

Clamp die

Multi-ball
mandrel

Cleat

Wiper die

Bend die

Cleated clamp plug

7.32 Schematic of the tooling used for bending tubes (Luo, 2008).

with heated tooling, as shown in Fig. 7.34 (Luo and Sachdev, 2005, pp. 477–482). In this bending process, three parts of the bend tooling, the bend die, the pressure die, and the mandrel, were heated, and the temperature for each heating zone was controlled separately. Using the optimal parameters developed for the moderate temperature bending process (Luo and Sachdev, 2005, pp. 477–482), a 2D/90° bend, as shown in Fig. 7.35, was consistently achieved with AZ31 and AM30 tubes at about 150 °C. No surface cracks were detected in these bent tubes.

In the cold rotary draw bending process, the tension (outer) side of the tube is thinned while the compression (inner) side is thickened. Figure 7.36 shows the degree of thinning along the tension side of magnesium alloy tubes bent to the 2D/90° condition at 150 °C. The smooth thinning distribution in Fig. 7.36 suggests uniform deformation during bending at elevated temperatures. The improved bendability of the magnesium alloy tubes is due to the higher ductility and formability reported in the literature (Luo and Sachdev, 2004, pp. 79–85; Krajewski, 2001, pp. 175–179).

7.4.4   Forming of magnesium extrusions

Elevated temperatures increase the formability of magnesium alloys. More complex geometries and greater circumferential expansion can be achieved at

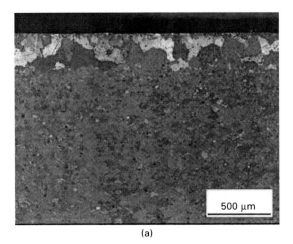

(a)

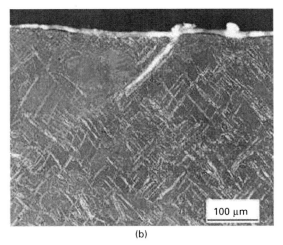

(b)

7.33 Optical micrographs showing the microstructure of an AZ31 tube: (a) before bending; and (b) after bending at room temperature (fractured at 30°) (Luo and Sachdev, 2008).

elevated temperatures using pressurized fluids such as oil or gas. Recently, a warm gas forming process at a much lower temperature range (150–350°C) was developed using an Interlaken 5000-KN press (Figure 7.37) (Martin *et al.*, 2005, pp. 51–56). This press is equipped with two separate controllers and pressure intensifiers to perform the tests with either gas at elevated temperature or water at room temperature. For warm forming, nitrogen is pre-charged into the system with a pre-charge control valve, and forming under gas pressure or volume control is achieved with a pressure intensifier. Fast forming cycles of about 10 seconds can be obtained with either operating mode. Two end-feed actuators are used to seal and provide axial feeding

Tube

Heated dies

Tooling
temperature
controller

Bender

7.34 A rotary draw bending machine with heated tooling for magnesium
tube bending.

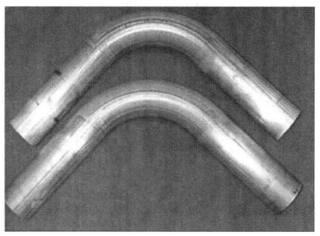

7.35 AM30 and AZ31 tubes bent (2D/90°) at 150 °C (Luo and Sachdev,
2008).

during forming. The forming-die holder is heated with electric resistance
elements and a temperature controller is used to measure and control the
temperatures independently in the six zones of the die. The die cavity has
two rectangular zones, a large and a small zone, and die inserts are used to
create different geometries required for tube forming. Complete expansion
involves an average tube expansion of 25 percent and 50 percent in the
smaller and larger cavity regions, respectively. Figure 7.38 compares an
AZ31 magnesium alloy tube hydroformed at room temperature with a similar
tube gas-formed at 250 °C. While both tubes failed at an extrusion seam,
gas forming could provide a circumferential expansion of 28% compared to
only 8% with hydroforming at room temperature. Further optimization of
extrusion and warm forming process would improve the formability of the

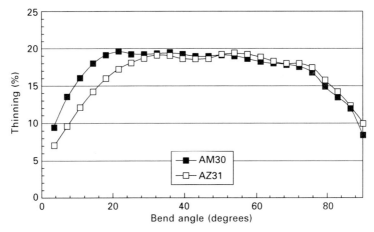

7.36 Thinning distribution in magnesium tubes bent at 150°C (Luo and Sachdev, 2008).

7.37 Warm gas forming press at CANMET, Ottawa, Canada (Martin *et al.*, 2005).

magnesium tubes, especially if controlled end feeding could be accomplished. Alternatively, seamless extrusions can be considered.

Seamless extrusions of AZ31 magnesium alloy were the subject of another study of warm hydroforming (gas formed using nitrogen) (Ben-

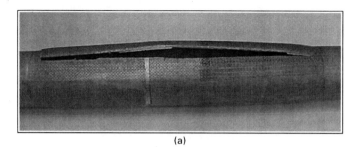

(a)

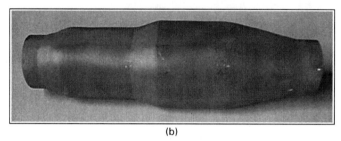

(b)

7.38 (a) Room temperature hydroforming; and (b) warm gas forming of magnesium alloy AZ31 tubes (Luo and Sachdev, 2008).

Artzy *et al.*, 2006, pp. 253–258). Figure 7.39(a) shows the schematic of the warm forming process. The tooling consists of two units, the clamping and the forming dies. The clamping unit ensures the sealing of the tube and the supply of the pressure medium. The tube is put into the forming unit, which consists of a main body and a free expansion guided zone equipped with cooling and heating devices. The guided zone makes it possible that tubes of different outer diameters can be formed under plane strain condition. Band heaters are used to heat the forming unit and the sealing punches. Figure 7.39(b) shows an AZ31 tube formed at 350°C with 80 percent circumference expansion.

7.4.5   Automotive applications of extruded magnesium

Magnesium extrusions are used in aerospace, nuclear, luggage, hand-tools, bicycle and motorcycle applications, but there have been no reported applications of magnesium extrusions in the automotive industry. Table 7.10 summarizes the potential applications of magnesium extrusions in automotive interior, and body and chassis areas, some of which may involve hydroforming processes.

Magnesium extrusions have been used to make prototype parts, such as bumper beams and most parts of a spaceframe for the VW 1-Liter Car (Schumann and Friedrich, 2003). However, the production of magnesium

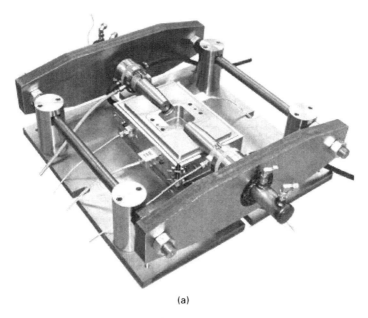

(a)

(b)

7.39 Warm gas forming of magnesium alloy tubes (a) lower die and (b) AZ31 tube formed at 350°C (80 per cent circumference expansion) (Ben-Artzy *et al.*, 2006) (used with permission from TMS).

Table 7.10 Potential automotive applications of magnesium extrusions

System	Component
Interior	Instrument panel
	Seat components
	HVAC (Heating, ventilation and air-conditioning) components
Body	Roof frame
	Bumper beam
	Radiator support
	Shotgun
	Frame rail
Chassis	Engine cradle
	Subframe

tubes/extrusions in automotive structures would require more development to meet all performance and cost targets as well as a supply base for high volume automotive production. Recently, a US–Canada–China collaborative effort has been focused on the development of enabling technologies for magnesium body applications using front end body structure as a test bed. A magnesium front end would weigh about 35 kg and provide up to 40 kg of mass saving, compared with the equivalent steel construction (Luo and McCune, 2008). Bending and gas forming processes of magnesium tubes will be further developed and validated in this project for automotive applications.

7.5   Future trends

While magnesium is the lightest structural metal and the third most commonly used metallic material in automobiles following steel and aluminum, many challenges remain in various aspects of alloy development and manufacturing processes to exploit its high strength-to-mass ratio for widespread lightweight applications in the automotive industry.

7.5.1   Material challenges

Compared with the numerous aluminum alloys and steel grades, there are only a limited number of low-cost magnesium alloys available for automotive applications. The conventional Mg-Al based cast alloys offer moderate mechanical properties due to the limited age-hardening response of this alloy system. Since the development of vacuum die casting and other high-integrity casting processes, magnesium castings can be heat-treated with no blisters. Alloy systems with significant precipitation hardening such as Mg-Sn (Mendis *et al.*, 2006, pp. 163–171; Luo and Sachdev, 2009, pp. 437–443) and Mg-rare earths (Fu *et al.*, 2008, pp. 182–192) should be developed with improved mechanical properties. New alloys with improved ductility, fatigue strength, creep resistance and corrosion resistance should also be explored.

For wrought magnesium alloys, the strong plastic anisotropy and tension–compression asymmetry due to texture remain obstacles for many structural applications. Microalloying with elements such as Ce (Mishra *et al.*, 2008, pp. 562–565), to improve the plasticity of magnesium alloys, has proven to be an effective approach in wrought magnesium alloy development. Other alloys systems such as Mg-Zn-rare earths have shown more 'isotropic' mechanical properties. Although additions of more than 11% Li can transform the hcp α-Mg solid solution into highly workable and more isotropic body-centered cubic alloys, automotive applications of this alloy will remain difficult.

The properties of magnesium alloys can be significantly enhanced if micro- and nano-particles are introduced to form metal matrix composites (MMC). Micro- and nano-sized particles offer strengthening mechanisms in different length scales and provide a tremendous opportunity for a new class of engineering materials with tailored properties and functionalities for automotive applications. Computational thermodynamics and kinetics will be used in the design and manufacture of these MMC materials.

7.5.2   Process challenges

The current success of magnesium is primarily attributed to its superior die-castability compared with aluminum alloys. However, more challenges exist in gravity, low-pressure and squeeze casting of magnesium alloys due to the need to compensate for larger shrinkage when compared with aluminum alloys. Melt handling, molten metal transfer, grain-refinement, and kind and amount of die lubrication coating as well as casting parameters need to be developed specifically for magnesium alloys to fully utilize their intrinsic properties.

Various forming processes need to be optimized for magnesium alloys. Elevated temperature forming is needed for most extrusion and sheet components. Research efforts have been directed to lowering the forming temperatures and reducing the cycle times. New forming processes should be developed to utilize the dramatically improved formability of magnesium alloys at certain ranges of temperature and strain rate. Room temperature (RT) or near-RT forming techniques are also being explored for new magnesium alloys such as Mg-Zn-Ce alloy (Bohlen, *et al.* 2007, pp. 2101–2112).

Fabrication of Mg-based MMCs includes liquid-mixing and preform infiltration. Both processes rely on the ability to achieve controlled dispersion (distribution) of the reinforcement phase in the metal matrix. Dispersion can be enhanced by new techniques such as ultrasonic, friction-stir and vacuum die casting for MMC fabrication.

Welding and other joining methods comprise the very important final step in assembling magnesium components and sub-systems. There is a wide range of joining techniques that can be potentially used for Mg-to-Mg, Mg-to-Al and Mg-to-steel joining. However, to date there have not been welded magnesium structures in automotive applications. Some of the techniques being developed for magnesium include arc-welding, laser welding, resistance spot welding (RSW), friction stir welding (FSW), adhesive bonding, self piercing rivets (SPR) and mechanical fastening (http://www.aist.go.jp/aist_e/latest_research/2008/20081105/20081105.html).

7.5.3   Performance challenges

There are several performance-related challenges that need significant research efforts. Some of them are highlighted in two current USAMP projects, the Magnesium Powertrain Cast Components Project (Powell, 2009; Beals *et al.*, 2007) and the Magnesium Front End Research and Development Project (Luo *et al.*, 2008, pp. 3–10).

Crashworthiness

Magnesium castings have been used in many automotive components such as the instrument panel beams and radiator support structures. High-ductility AM50 or AM60 alloys are used in these applications and performed well in crash simulation and tests; and many vehicles, with these magnesium components achieved five-star crash rating. However, there is limited material performance data available for component design and crash simulation. A recent study shows that magnesium alloys can absorb significantly more energy than either aluminum or steel on an equivalent mass basis (Easton *et al.*, 2008, pp. 57–62). While steel and aluminum tubes fail by progressive folding in crash loading (more desirable situation), magnesium alloy tubes (AZ31 and AM30) tend to fail by sharding or segment fracture (Easton *et al.*, 2008, pp. 57–62; Wagner *et al.*, 2009). However, the precise fracture mechanisms for magnesium under crash loading are still not clear, and material models for magnesium fracture are needed for crash simulation involving magnesium components. Additionally, new magnesium alloys need to be developed to have progressive folding deformation in crash loading.

Noise, vibration and harshness (NVH)

It is well known that magnesium has high damping capability, but this can be translated into better NHV performance only for mid range sound frequency; 100–1000 Hz (Logan *et al.*, 2006). Low-frequency (<100 Hz) structure-borne noise can be controlled by component stiffness between the source and receiver of the sound. The lower modulus of magnesium, compared to steel, is often compensated by thicker gages and/or ribbing designs. For high-frequency (>1000 Hz) airborne noise, a lightweight panel, regardless of material, would transmit significantly more road and engine noise into the occupant compartment unless the acoustic frequencies could be broken up and damped. Magnesium, with its low density, is disadvantaged for this type of applications unless new materials with laminated structures are developed for sound insulation; greater damping.

Fatigue and Durability

Fatigue and durability are critical in magnesium structural applications and there is limited data in the literature, especially on wrought alloys. The effect of alloy chemistry, processing and microstructure on the fatigue characteristics of magnesium alloys need to be studied. Casting, extrusion and sheet products needs to be characterized sufficiently to establish links between microstructural features and fatigue behavior. Multi-scale simulation tools can be used to predict the fatigue life of magnesium components and sub-systems, which can be validated for automotive applications.

Corrosion

Pure magnesium has the highest standard reduction potential of the structural automotive metals (see Table 7.11). As noted earlier, while pure magnesium (at least with very low levels of iron, nickel, and copper) has atmospheric corrosion rates that are similar to those of aluminum, magnesium's high reduction potential makes it very susceptible to galvanic corrosion; that is, when magnesium is in electrical contact with other metals below it in the reduction potential table. The impact of this susceptibility to galvanic

Table 7.11 Standard reduction potential of common metals

Electrode	Reaction	Potential (V)
Li, Li$^+$	Li$^+$ + e$^-$ $\Rightarrow$ Li	−3.02
K, K$^+$	K$^+$ + e$^-$ $\Rightarrow$ K	−2.92
Na, Na$^+$	Na$^+$ + e$^-$ $\Rightarrow$ Na	−2.71
Mg, Mg^{2+}	Mg^{2+} + 2e$^-$ $\Rightarrow$ Mg	−2.37
Al, Al^{3+}	Al^{3+} + 3e$^-$ $\Rightarrow$ Al	−1.71
Zn, Zn^{2+}	Zn^{2+} + 2e$^-$ $\Rightarrow$ Zn	−0.76
Fe, Fe^{2+}	Fe^{2+} + 2e$^-$ $\Rightarrow$ Fe	−0.44
Cd, Cd^{2+}	Cd^{2+} + 2e$^-$ $\Rightarrow$ Cd	−0.40
Ni, Ni^{2+}	Ni^{2+} + 2e$^-$ $\Rightarrow$ Ni	−0.24
Sn, Sn^{2+}	Sn^{2+} + 2e$^-$ $\Rightarrow$ Sn	−0.14
Cu, Cu^{2+}	Cu^{2+} + 2e$^-$ $\Rightarrow$ Cu	0.34
Ag, Ag$^+$	Ag$^+$ + e$^-$ $\Rightarrow$ Ag	0.80

Source: Hawke *et al.*, 1999.

corrosion on the application of magnesium in exposed environments is severe in both the macro-environment and the micro-environment. In the macro-environment, magnesium alloys must be electrically isolated from other metals to prevent the creation of galvanic couples; e.g. the use of steel bolts in direct contact with magnesium. Isolation can be achieved by replacing the bolt with a less reactive metal, as has been done in the Mercedes automotive transmission case where steel bolts have been replaced with aluminum bolts, (Greiner *et al.*, 2004). Isolation can also be achieved by coating the 'other' metal. Finally, isolation can be achieved by the use of shims or spacers of compatible materials of sufficient geometry and size to prevent electrical contact in the presence of salt water, as shown, for example, for the Corvette cradle (Fig. 7.16). While the casting cost can be competitive with aluminum, the isolation strategies required can often make the application more expensive and thus restrictive in its use. Much work is being conducted to address corrosion issues with magnesium but these are beyond the scope of this chapter (Song, 2005).

A major challenge in magnesium automotive applications is to establish the surface finishing and corrosion protection processes. The challenge is two-fold since surface treatments to magnesium play roles in both manufacturing processes (e.g. adhesive bonding) as well as in the product life cycle that demands corrosion resistance. Furthermore, the current manufacturing paradigm for steel-intensive body structures employs chemistries in the paint shop that are corrosive to magnesium. Conditions are further aggravated by galvanic couples containing primarily steel fasteners. Future research will explore novel coating and surface treatment technologies including pretreatments such as micro-arc anodizing, non-chromated conversion coatings, and 'cold' metal spraying of aluminum onto magnesium surfaces. Since most studies of corrosion protection and pre-treatment of magnesium have focused on die castings, the behavior of sheet, extrusion and high-integrity castings will be explored for process compatibility.

In summary, the future success of magnesium as a major automotive material will depend on how the technical challenges are addressed. These challenges are huge and global, and would require significant collaboration among industries, governments and academia from many countries. It is very encouraging that many of these international and interdisciplinary collaborations are being nurtured for magnesium and it is expected that future developments will enable us further utilize the benefits of magnesium, the lightest structural metal.

7.6   Acknowledgments

The authors gratefully acknowledge the careful read and insightful comments by Dr. Anil K. Sachdev, General Motors Research & Development Center.

7.7   References

Aghion E, Moscovitch N, Arnon A, and Bronfin B (2007), 'The Capability of MRI Magnesium Alloys to Address High Temperature Application Requirements', in Beals R, Luo A, Neelameggham N, and Pekguleryuz M, eds., *Magnesium Technology 2007*, TMS, Warrendale, Pennsylvania, 253–255.

ASM International (1988), *Metals Handbook*, **15**, ASM International, Materials Park, Ohio.

ASM International (1992), *Metals Handbook*, **2**, ASM International, Materials Park, Ohio.

ASM International (1998), David J R, *Metals Handbook, Desk Edition*, 2nd Edition, ASM International, Materials Park, Ohio.

ASM International (1999), in Avedesian M and Baker H, eds., *Magnesium and Magnesium Alloys*, ASM International, Materials Park, Ohio.

ASTM International, Committee B07 on Light Metals and Alloys (2008), *Standard Practice for Codification of Unalloyed Magnesium and Magnesium – Alloys, Cast and Wrought'*, ASTM International Designation B951-08, ASM International, West Conshohocken, Pennsylvania.

Barnes A J (2002), 'Industrial Applications of Superplastic Forming', *23rd Annual Aluminum Design and Fabrication Seminar*, October 23, 2002, Livonia, Michigan.

Barnes A J (2007), 'Superplastic Forming 40 Years and Still Growing,' *Journal of Materials Engineering and Performance*, **16**, 440–454.

Barnes L T (1992), 'Rolled Magnesium Products: What Goes Around, Comes Around', *Proceedings of the International Magnesium Association*, Chicago, Illinois, 29–43.

Beals R S, Liu Z-K, Jones J W, Mallick P K, Emadi D, Schwam D, and Powell B R (2007), 'US Automotive Materials Partnership, Magnesium Powertrain Cast Components Project: Fundamental Research Summary', *Journal of Minerals, Metals, and Materials*, **59** (8) 43–8.

Behrens B-A and Hubner S (2005), 'Heated Clinching of Magnesium Sheet Metal', *Annals of the German Academic Society for Production Engineering*, Production Engineering Research Development, **XII** (1), 59–62.

Ben-Artzy A, Spinat E, Dahan O, Siegert K, Jager S, Mueller K, Altan T (2006), 'High Internal Pressure Forming of Magnesium Tubes', in Luo A A, Neelameggham N R and Beals R S, eds. *Magnesium Technology 2006*, TMS, Warrendale, Pennsylvania, 253–258.

Bohlen J, Nuernberg M, Senn J W, Letzig D, and Agnew S R (2007), 'The Texture and Anisotropy of Magnesium-Zinc-Rare Earth Alloy Sheets', *Acta Materialia*, **55** (6), 2101–2112.

Brown, R E (2008), 'Future of Magnesium Developments in 21st Century', presentation at Materials Science & Technology Conference, Pittsburgh, Pennsylvania, USA, October 5–9.

Brown Z, *et al.* (2009), 'Development of Super-vacuum Die Casting Process for Magnesium Alloys', to be published in *Cast Expo '09 Proceedings*, North American Die Casting Association, Wheeling, Illinois.

Buschow K H J and van Vucht J H N (1967), 'Systematic Arrangement of the Binary Rare earth–Aluminum Systems', *Philips Research Reports*, **22,** 233–45.

Busk R S (1987), *Magnesium Product Design*, Marcel Dekker, Inc., New York and Basel.

Bussy A, *Mémoire sur le Radical métallique de la Magnésie*, 1831.

Carpenter J A, Jackman J, Osborne R J, Powell B R, Li N, and Sklad P (2008), 'Automotive Research and Development in North America,' *Die Casting Engineer*, **52** (3), 54–9.

Carsley J E and Kim S (2007), 'Warm Hemming of Magnesium Sheet,' *Journal of Materials Engineering and Performance*, **16**, 331–338.

Carter J, Krajewski P E, and Verma R (2008), 'The Hot Blow Forming of AZ31 Mg Sheet: Formability Assessment and Application Development', *Journal of Minerals, Metals, and Materials*, **60** (11), 77–81.

DiFrancesco C A and Kramer D A (2008), 'Magnesium Metal Statistics', U.S. Geological Survey. Available from http://minerals.usgs.gov/minerals/pubs/commodity/magnesium/

Easton M, Beer A, Barnett M, Davies C, Dunlop G, Durandet Y, Blacket S, Hilditch T, and Beggs P (2008), 'Magnesium Alloy Applications in Automotive Structures', *Journal of Minerals, Metals, and Materials*, **60** (11), 57–62.

Emley E (1966), *Principles of Magnesium Technology*, 1st Ed., Pergamon Press, Ltd., Oxford, England, 828.

Flemings M C (1974), *Solidification Processing* (Materials Science and Engineering Series), McGraw-Hill College, New York.

Friedrich H and Schumann S (2001), 'Research for a "New Age of Magnesium" in the Automotive Industry', *Journal of Materials Processing Technology*, **117**, 276–281.

Fu P, Peng L, Jiang H, Chang J, and Zhai C (2008), 'Effects of Heat Treatments on the Microstructures and Mechanical Properties of Mg–3Nd–0.2Zn–0.4Zr (wt.%) Alloy', *Materials Science and Engineering A*, **486**, 183–192.

Greiner J, Doerr C, Nauerz H, and Graeve M (2004), 'The New "7G-TRONIC" of Mercedes-Benz: Innovative Transmission Technology for Better Driving Performance, Comfort, and Fuel Economy', *SAE Technical Paper No. 2004-01-0649*, SAE International, Warrendale, Pennsylvania.

Groche R, Huber R, Dorr J, and Schmoeckel D (2002),'Hydromechanical Deep Drawing of Aluminium Alloys at Elevated Temperatures', *Annals of the College International Pour la Recherhe en Productique*, **51/1**, 215–218.

Habashi F (2006), 'A History of Magnesium', in Pekguleryuz M and Mackenzie L, eds., *Magnesium Technology in the Golden Age*, Montreal, METSOC, 31–42.

Hawke D L, Hillis J E, Pekguleryuz M, and Nakatsugawa I (1999), 'Corrosion Behavior' in Avedesian M M and Baker H, eds, *Magnesium and Magnesium Alloys*, ASM International, Materials Park, Ohio, 194.

Hillis J (1983), 'The Effects of Heavy Metal Contamination on Magnesium Corrosion Performance', *SAE Technical Paper No. 830523*, SAE International, Warrendale, Pennsylvania.

Hoeschl M, Wagener W, and Wolfe J (2006), 'BMW's Magnesium–Aluminum Composite Crankcase: State-of-the-Art Light Metal Casting and Manufacturing', *SAE Technical Paper No. 2006-01-0069*, SAE International, Warrendale, Pennsylvania.

Hollrigl-Rosta F (1980), 'Magnesium in Volkswagen', *Light Metal Age*, **8**, 22–29.

Hollrigl-Rosta F, Just R, Kohler J, and Melzer, H-J (1980), 'Magnesium in Volkswagen', *Metall*, **12**, 12.

Jorstad J (1993), *Chemical and Physical Characterisations of Molten Aluminium*, in 'Aluminum Casting Technology', 2nd Edn, ed. Zalensas D, The American Foundrymen's Society, Des Plaines, Illinois, 21.

Kainer K and Benzler T (2003), 'Squeeze-casting and Thixo-casting of Magnesium Alloys', in Kainer K *Magnesium – Alloys and Technologies*, Wiley-VCH Verlag GmbH & Co. KGaA, Weinheim, Germany, 58.

Kaye A and Street A (1982), *Die Casting Metallurgy*, Butterworth Scientific, London.

Krajewski P E (2001), 'Elevated Temperature Forming of Sheet Magnesium Alloys', *SAE International Technical Paper No. 2001-01-3104*, Warrendale, Pennsylvania, 175–179.

Krajewski P E and Schroth J G (2007), 'Overview of Quick Plastic Forming Technology', *Materials Science Forum*, **551/552**, 3–12.

Krajewski P E, *et al.* (2006), 'Warm Forming of Aluminum: Summary of USAMP Project AMD307', presentation at Materials Science & Technology Conference, Oct. 15, Cincinnati, Ohio.

Krajewski P E, Kim S, Carter J T, and Verma R (2007), 'Magnesium Sheet: Automotive Applications and Future Opportunities', *Trends in Metals and Materials Engineering*, **20** (5), 60–68.

Kunst M *et al.* (2006), 'Creep Deformation and Mechanisms of AJ (Mg–Al–Sr) Alloys', in Pekguleryuz M and Mackenzie L, eds., *Magnesium Technology in the Golden Age*, Montreal, METSOC, 647–661.

LeBeau S E, Yamamoto Y, and Sakamoto K (1998), 'Thixomolding of Magnesium Automotive Components', *SAE International Technical Paper No. 980087*, SAE International, Warrendale, Pennsylvania.

Li N, Osborne R, Cox B, and Penrod D (2005), 'Magnesium Engine Cradle – The USCAR Structural Cast Magnesium Development Project', *SAE International Technical Paper No. 2005–01–0337*, SAE International, Warrendale, Pennsylvania.

Logan S, Kizyma A, Patterson C, and Rama S (2006), 'Lightweight Magnesium-intensive Body, Structure', *SAE International Technical Paper No. 2006-01-0523*, SAE International, Warrendale, Pennsylvania.

Luo A A and Sachdev A K (2004), 'Mechanical Properties and Microstructure of AZ31 Magnesium Alloy Tubes', in Luo A A, *Magnesium Technology 2004*, ed. The Minerals, Metals and Materials Society (TMS), Warrendale, Pennsylvania, 79–85.

Luo A A and McCune R C (2008), 'Magnesium Front End Projects', V.S. Dept. of Energy, *Automotive Lightweighting Materials*, FY 2006 Progress Report, TMS 2008.

Luo A A and Sachdev A K (2005), 'Development of a Moderate Temperature Bending Process for Magnesium Alloy Extrusions', presented at The International Conference on Magnesium, Beijing, China, Sept. 20–24, 2004, and published in *Materials Science Forum*, **488/489**, 477–482.

Luo A A and Sachdev A K (2007), 'AM30 – A New Wrought Magnesium Alloy', in Beals R S, Luo A A, Neelameggham N R, and Pekguleryuz M O, *Magnesium Technology 2007*, TMS, Warrendale, Pennsylvania, 321–326.

Luo A A and Sachdev A K (2008), 'Bending and Hydroforming of Aluminum and Magnesium Alloy Tubes', in Koc M, ed., *Hydroforming for Advanced Manufacturing*, Woodhead Publishing, Cambridge, England.

Luo A A and Sachdev A A (2009), 'Microstructure and Mechanical Properties of Mg–Al–Mn and Mg–Al–Sn Alloys', in E.A. Nyberg, S.R. Agnew, N.R. Neelameggham and M.O. Pekguleryuz, eds. *Magnesium Technology 2009*, TMS, Warrendale, Pennsylvania, 437–443.

Luo A A *et al.* (2001), 'Creep and Microstructure of Magnesium–Aluminum-Calcium-based Alloys, *Metallurgical and Materials Transactions*, **33A**, 567–574.

Luo A A, Sachdev A K, Mishra R K and Kubic R C (2005), 'Bendability and Microstructure of Magnesium Alloy Tubes at Room and Elevated Temperatures', in Neelameggham N R, Kaplan H I, and Powell B R, eds. *Magnesium Technology 2005*, TMS, Warrendale, Pennsylvania, 145–148.

Luo A A, Nyberg E A, Sadayappan K, and Shi W (2008), 'Magnesium Front End Research and Development: A Canada–China–USA Collaboration', in M.O. Pekguleryuz, N.R. Neelameggham, R.S. Beals and E.A. Nyberg, eds. *Magnesium Technology 2008*, TMS, Warrendale, Pennsylvania, 3–10.

Margrave J (1967), *The Characterization of High Temperature Vapors*, John Wiley & Sons, New York, New York.

Martin P, Baragar D, Boyle K P, Luo A A, Jonas J J, Godet S, Neale K W (2005), 'Elevated Temperature Property Measurements for Warm Forming Aluminium Alloy Tubes', in *Proceedings of the 2nd International Light Metals Technology Conference*, ed. Kaufmann H, June 8–10, 2005, St. Wolfgang, Austria. 51–56.

Mendis C L, Bettles C J, Gibson M A, and Hutchinson C R (2006), 'An Enhanced Age Hardening Response in Mg–Sn Based Alloys Containing Zn', *Materials Science and Engineering A* **435/436**, 163–171.

Mercer W E (1990), 'Magnesium Die Cast Alloys for Elevated Temperature Applications', *SAE Technical Paper No. 900788*, SAE International Warrendale, Pennsylvania.

Mishra R K, Gupta A K, Rao P R, Sachdev A K, Kumar A M, and Luo A A (2008), 'Influence of Cerium on the Texture and Ductility of Magnesium Extrusions', *Scripta Materialia*, **59**, 562–565.

Moll F, Mekkaoui F, Schumann S, and Friedrich H (2004) 'Application of Mg sheets in Car Body Structure', in Kainer K U Ed. *Magnesium: Proceedings of the 6th International Conference Magnesium Alloys and Their Applications*, Wiley-Vch, Weinheim, Germany.

Nelson K E (1970), 'Magnesium Die Casting Alloys', *SDCE Transactions*, Paper No. 13.

Okamoto H (1988), 'Mg (Magnesium)', in A.A. Nayeb-Hashema and J.B. Clark, eds., *Phase Diagrams of Binary Magnesium Alloys*, ASM International, 1–3.

Pidgeon L M, Mathes J C, Woldman N E, Winkler J V, and Loose W S (1946), *Magnesium*, ASM International, 4–22.

Polmear I J (1999), in M.M. Avedesian and H. Baker, eds., *Magnesium and Magnesium Alloys*, ASM International, 3–5.

Powell B R (2009), 'Magnesium Powertrain Cast Components,' in *FY2008 Annual Progress Report for Automotive Lightweighting Materials*, U.S. Department of Energy, Washington, D.C., April.

Powell B R, Luo A A, Rezhets V, Bommarito J J, and Tiwari B L (2001), 'Development of Creep-Resistant Magnesium Alloys for Powertrain Applications: Part 1 of 2', *SAE Technical Paper 2001-01-0422*, SAE International, Warrendale, Pennsylvania.

Powell B R, Rezhets V, Balogh M P, and Waldo R A (2002), 'Microstructure and Creep Behavior in AE42 Magnesium Die-Casting Alloy', *TMS Journal of Minerals, Metals, and Materials*, **54** (8), 34–38.

Powell B, *et al.* (2005), 'Progress Toward a Magnesium-intensive Engine: the USAMP Magnesium Powertrain Cast Components Project', *SAE 2004 Transactions, Journal of Materials & Manufacturing Paper No. 2004-1-0654*, SAE International, Warrendale, Pennsylvania, 250–259.

Rashid M S *et al.* (2001), *Quick Plastic Forming of Aluminum Alloy Sheet Metal*. US Patent 6 253 588, July 3, 2001.

Savage K and King J F (2000), 'Hydrostatic Extrusion of Magnesium', in Kainer K U, ed. *Magnesium Alloys and Their Applications*, Wiley-VCH, Weinheim, Germany, 609–614.

Schroth J G (2004), in Taleff E M ed. *Advances in Superplasticity and Superplastic Forming* TMS, Warrendale, Pennsylvania.

Schumann S and Friedrich H (1998), 'The Use of Magnesium in Cars – Today and in the Future,' in Mordike B and Kainer K, eds., *Magnesium Alloys and their Applications*, Werkstoff-Informationsgesellschaft, Frankfurt, Germany, 3–13.

Schumann S and Friedrich H (2003), 'Current and Future Use of Magnesium in the Automobile Industry', *Magnesium Alloys 2003*, Materials Science Forum, Trans Tech, Switzerland, **419–422**, 51–56.

Sieracki E G, Velazquez J J, and Kabiri K (1996), 'Compressive Stress Retention Characteristics of High Pressure Die Casting Magnesium Alloys', *SAE Technical Publication No. 960421*, SAE International, Warrendale, Pennsylvania.

Song G (2005), 'Recent Progress in Corrosion and Protection of Magnesium Alloys', *Advanced Engineering Materials*, **7**, 7, 563–586.

Taub A I, Krajewski P E, Luo A A, and Owens J N (2007), 'The Evolution of Technology for Materials Processing over the Last 50 Years: The Automotive Example', *Journal of Minerals, Metals, and Materials*, **59** (2) 48–57.

Timminco Corporation (1998), *Magnesium Wrought Products*, Timminco Corporation Brochure, Aurora, Colorado.

Verma R and Carter J T (2006), 'Quick Plastic Forming of a Decklid Inner Panel with Commercial AZ31 Magnesium Sheet', *SAE International Technical Paper No. 2006-01-0525*, SAE International, Warrendale, Pennsylvania.

Vinarcik E (2003), *High Integrity Die Casting Processes*, John Wiley & Sons, Hoboken, New Jersey.

Wagner D A, Logan S D, Wang K, Skszek T, and Salisbury, C P (2009), 'Test Results and FEA Predictions from Magnesium AM30 Extruded Beams in Bending and Axial Compression', in E.A. Nyberg, S.R. Agnew, N.R. Neelameggham and M.O. Pekguleryuz, eds. *Magnesium Technology 2009*, TMS, Warrendale, Pennsylvania.

Webb L (2003), 'Magnesium Supply and Demand 2002', in *International Magnesium Association 60th Annual World Magnesium Conference*, Washington, D.C., 31–34.

Wei L Y and Dunlop GL (1992), 'Precipitation Hardening in a Cast Mg–Rare Earth Alloy', in *Magnesium Alloys and Their Applications*, DGM Informationsgesellschaft, Verlag, Germany, 335.

8
Polymer and composite moulding technologies for automotive applications

P. MITSCHANG and K. HILDEBRANDT, Institut für
Verbundwerkstoffe GmbH, Germany

Abstract: This chapter examines polymer and composite moulding technologies used for automotive applications. It is divided into several sections starting with an overview on which polymeric materials and composite processing technologies are used nowadays. Processing technologies focus on liquid composite moulding (LCM), sheet/bulk moulding compounds (SMC/BMC), long fibre-reinforced thermoplastics and glass fibre mat thermoplastics (LFT/GMT), thermoforming and injection moulding. The following sections comprise different examples for fibre-reinforced polymer composites (FRPC) in interior and exterior, chassis and body applications. The chapter closes by looking at further challenges for composites in the automotive industry.

Key words: polymer composites, exterior composite components, liquid composite moulding, thermoforming.

8.1   Introduction

Within the last few decades polymers and polymer composites have gained more and more importance for the automotive industry. To summarise the success the following chapter gives the reader an overview on how polymers and polymer composites are used in the automotive industry nowadays and in the past. The chapter is divided into several sections starting with an overview of what kinds of polymers are used including both non-reinforced materials and fibre-reinforced materials. Section 8.3 describes the different processing technologies for manufacturing polymer composite materials. It is divided into four parts starting with liquid composite moulding and compression moulding techniques used for thermosets. Long fibre thermoplastics are widely used in the automotive industry and are processed using either compression moulding or injection moulding techniques, which are also described in this section. The following part describes the thermoforming of endless fibre-reinforced thermoplastics while the last part deals with the injection moulding technique. Section 8.4 describes examples for interior applications, chassis and exterior applications and body applications, to give the reader an overview on the applications of composites found on the market. The last section gives an outlook on future challenges for polymer composites.

210

8.2   Polymeric materials used in the automotive industry

In general, polymers can be classified in three categories, thermoplastics, thermosets and elastomers. All three polymer classes are built-up using monomers as their smallest component and differ mainly in the number of crosslinks between the distinct polymer chains and their molecular structure. Thermoplastics show the lowest crosslink density leading to remeltable polymers. Thermosets on the other hand have a very high crosslink density which leads to non-meltable polymerswith a three-dimensional polymer network. Elastomers usually have a medium crosslink density and a glass transition temperature (T_g) considerably below room temperature which leads to non-meltable but very elastic polymers (see Fig. 8.1). The specific chemical and physical properties for each polymer class therefore lead to unique manufacturing processes applicable for each class.

8.2.1   Non-reinforced standard materials

Non-reinforced standard polymers comprise all neat polymers and polymers with particle reinforcement (e.g. short and long glass fibres, chalk, carbon

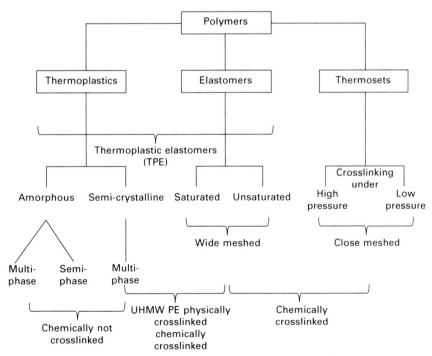

8.1 Classification of polymers into thermoplastics, thermosets and elastomers.

black) used in the automotive industry. As required properties are very variable, a wide range of polymers are used in automotive products. Table 8.1 shows an overview of the standard polymers used. The table makes no claim to be complete, especially, because in recent years the use of polymer blending has brought a vast number of combinations on to the market, which try to combine the desired properties of each polymer in a synergistic way. Nevertheless, the first six polymers shown (PP, PE, PA, ABS, PUR and PVC) cover about 80% of the total polymer consumption in the automotive industry (VKE, 1999).

Table 8.1 Thermoplastics used in the automotive industry

Polymer	Abbreviation	Properties	Sample applications
Polypropylene	PP	Low cost, high strength, chemically resistant	Bumper, wheel housing, air filter housing
Polyethylene	PE	Low cost, chemically resistant, high strength, non-ageing	Fuel tank
Polyamide	PA	Low gas diffusivity, temperature-stable, high fatigue strength and stiffness, non-ageing	Engine cover, plug connector, air duct
Acrylonitrile-butadienestyrene	ABS	Dimension-stable, impact-resistant, high strength	Radiator grill, wheel flashing, interior panel
Polyurethane	PUR	Good damping behaviour, good electricity, low thermal conductivity	Seat cushion, instrument panel cushion, roof lining
Polyvinylchloride	PVC	Low cost, weatherproof, flame resistant	Underbody skidplate, wire insulation
Polyoxymethylene	POM	Chemically resistant, low creep, impact-resistant, temperature-stable, abrasion-resistant	Fastening clip, plug connector, cog wheel, bearing components
Poly(methyl-metacrylate)	PMMA	Transparent, UV-resistant, stress crack resistant, scratch resistant	Headlight lenses
Polycarbonate	PC	Transparent, UV-resistant, impact-resistant	Headlight lenses, car body parts
Polyethylene-terephthalate	PET	High tensile strength and stiffness, good barrier effect	Textile, safety seatbelt, airbag
Polybutylene-terephthalate	PBT	High stiffness, high heat deflection temperature (HDT), good electrical isolation, low coefficient of thermal expansion (CTE)	Electric housings, exterior mirror housing, handgrip

In addition to thermoplastics, thermosets are often used for fibre-reinforced composites.The resins mainly used are: polyester resins, epoxy resins and polyurethane resins.

8.2.2   Reinforcements

Reinforcements are often used to enhance the properties of polymers. These can be supplied in a variety of different shapes ranging from globular particles, to short and long fibres in the range of a few hundred microns to about 25 mm, up to endless fibres. Many applications produced by injection moulding or extrusion processes use reinforcements for various reasons. One of these reasons is the low price of reinforcement compared to the polymer price combined with enhanced mechanical properties such as strength and stiffness. Common filler materials are chalk, carbon black, chopped fibres and nanoparticles like TiO_2, SiO_2 or carbon nanotubes.

In endless fibre-reinforced composites the main task of the reinforcement is to enhance the mechanical property profile. Table 8.2 gives an overview on the most common endless fibre-reinforcements. It can be seen that depending on the required properties (modulus, strength or temperature) a large variety is available on the market.The standard reinforcements for endless fibre-reinforced composites are glass and carbon fibres. However, the market for natural fibres in the German automotive market in 2005, for example, was about 19 000 tons (without wood and cotton) with flax as the mostly used natural fibre (Karus *et al.*, 2006). In literature certain varieties of flax have been reported having ultimate tensile strengths and Young's moduli of up to 1500 MPa and 80 GPa respectively (Bhattacharyya and Fakirov, 2007). These promising properties combined with the eco friendliness of natural fibres may lead to ongoing market growth.

Table 8.2 Overview of endless fibre-reinforcements

Fibre		Tensile			Density (g/cm³)	T_{max} (°C)
		Young's modulus (GPa)	Strength (GPa)	Failure strain (%)		
Steel		200	2.8	4–20	7.8	1000
E-glass		80	3.5	4.0–5.4	2.54	300–350
SiC		400	4.8	0.9	2.8	1300
Carbon	PAN-HT	240	3.75	1.5–2.2	1.78	500
	PAN-HM	400	2.45	0.7–1.4	1.85	600
Aramid	Kevlar 49	135	3.5	2.5–2.7	1.45	250–300
UHMW-PE		172	3.3	4.0	0.97	100
Hemp		70	0.6	1.6	1.45	200
Flax		30	0.75	2.0–3.2	1.48	200
Sisal		20	0.6	2.0–2.5	1.45	200

Single fibres are most often prearranged into structures making them easier to handle and ready for further use. The first level of assembly is to create rovings from single fibres. These rovings are often assembled to two-dimensional structures as non-woven fabrics, woven fabrics and tapes. The decision which form to use is set by the pros and cons of each type taking into account the processability, stiffness, strength, interlaminar fracture toughness and cost of material. Further information can be found in the comprehensive work of Chou and Ko (1989).

8.2.3   Fibre-reinforced polymers

In this chapter the term fibre-reinforced polymer composite (FRPC) comprises only materials with an endless fibre reinforcement. All polymers using short and long fibre reinforcements are excluded and are dealt with separately. In general FRPCs consist of a polymeric matrix in which aligned endless fibres are embedded. The composite is then defined by the matrix, the reinforcement and the interface between the polymeric matrix and fibre reinforcement. The matrix consists either of a thermoplastic or a thermoset while the reinforcement is made of one of the materials listed in Table 8.1 and Section 8.2. By using various configurations of matrix, reinforcement, structure of the reinforcement, ratio between matrix and reinforcement and so on, one can obtain a vast number of different FRPCs each having a unique set of properties. The limits of these opportunities are usually set by the given price range or the manufacturing process.

8.3   Composite processing procedures

Composite manufacturing processes are divided in two major sections. The first section comprises the techniques for the manufacturing of thermoset material and the second section those processes used for the manufacturing of thermoplastic material. This division is mainly caused by the much higher viscosities of thermoplastics compared to thermosets. The most common process technologies that are used in the automotive industry are shown below.

8.3.1   Liquid composite moulding

The term liquid composite moulding (LCM) summarises a variety of process technologies such as resin transfer moulding (RTM), reaction injection moulding (RIM) and vacuum-assisted resin injection (VARI) that use thermoset resins and continuous fabric reinforcements to produce fibre-reinforced composites. Developed in the aviation industry, where the processing of prepreg materials in autoclaves is a very time consuming and expensive, liquid composite moulding allows a fast and high quality production of advanced composites. LCM processes are widely used for motor racing applications like Formula 1, but are currently evolving into suitable processes for mass markets. A general process cycle is as follows: a textile reinforcement is placed into a tool which is closed afterwards. The resin is stored in a separate reservoir. After the mould is closed, the resin is injected into the mould cavity, infiltrating the textile reinforcement. Typical processing viscosities of resins are in the range of 1–100 mPas. The time needed for infiltration varies from a few seconds for smaller parts up to hours for very large parts like rotor blades used for wind energy plants. The infiltration itself is a complex flow mechanism that combines macro and micro impregnation of the dry fibres. The polymerisation of the resin takes place in the mould, and is induced either by process heat or a chemical initiator. After the curing of the resin the part is demoulded.

8.3.2   Compression moulding

Compression moulding processes offer a high degree of automation, short cycle times, good reproducibility and excellent dimensional stability for both thermoplastics and thermosets which is the reason for diverse applications in various industrial sectors including the automotive industry. Thermoplastic compression moulding techniques include the processing of long fibre-reinforced thermoplastics (LFTs) and glass mat reinforced thermoplastics (GMTs), as well as the thermoforming of continuous organic sheets and films, while thermoset compression moulding is described by the processing of sheet moulding compounds (SMCs) and bulk moulding compounds (BMCs).

8.3.3   SMCs/BMCs

The processing of SMCs and BMCs can be divided into the process steps: preparation of the semi-finished part, transfer of the two-dimensional semi-finished part into the heated tool, press process and demoulding of the finished part.

The tool temperature is set at approximately 140 °C to 160 °C for unsaturated polyesters, which are the most widely used thermosets for sheet moulding compounds. The semi-finished parts are supplied as sheets and are assembled on shape and volume. During the press process the heated tool introduces energy into the thermoset which leads to a significant viscosity drop that combined with the pressure capacity forces the resin to flow and fill the mould cavity. As the curing process is a function of temperature and time, the viscosity of the resin increases during the press process and is the dominant factor influencing the overall process time.

The textiles used to reinforce the BMCs/SMCs are mostly random fibre mats with a typical fibre length of 25–50 mm. Recent developments utilise the lightweight potential by using aligned carbon fibre-reinforced SMC (C-SMC) (AVK, 2004). SMC has shown to be an excellent material for exterior body parts. To fulfil the requirements for class-A surfaces several components like mineral fillers and thermoplastic powders have to be added to matrix and reinforcement.

8.3.4   LFTs/GMTs

LFTs and GMTs are typically processed by a press process that can be divided into the following steps. If working with GMT material the first step is to assemble the GMT semi-finished part in shape and volume. The next step is to plasticise the LFT/GMT up to a molten state. LFT material consists of polymer pellets which are heated and converted into a composite melt using an extruder, while GMT material is put in an oven and heated using radiant heat and convection. The molten material is transported and put into the mould cavity. For polypropylene the temperature of the mould is typically in the range of 25 °C up to 80 °C, although nearly all polymers can be processed. The moment the semi-finished part touches the mould's surface it starts solidifying. Therefore, a fast closing press is necessary to avoid freezing of large amounts of polymer during the deposition of the LFT/GMT. Contrary to SMC/BMC processes the molten matrix flows from the centre of semi-finished material towards the edges. Due to its high flowability in the molten state, LFT allows a high degree of geometric freedom (Davis *et al.*, 2003).

Table 8.3 summarises and compares the different processes SMC, GMT and LFT.

Table 8.3 Comparison of SMC, GMT and LFT processes

	SMC	GMT	LFT
Used materials			
Matrix	Thermosets (e.g. UP-resin)	Thermoplastics (PP, PA, PET)	Thermoplastics (PP, PA, PET)
Reinforcement	Fibre mats	Glass fibre mats	Long glass fibres
Process parameters			
Tool temperature	140–160 °C	25–80 °C	25–80 °C
Pressure	50–100 bar	100–300 bar	100–300 bar
Cycle time	1 minute	30–40 seconds	30–40 seconds
Specific properties	Class-A, high volume production ability, excellent part reproducibility	Not suitable for visible parts, good impact energy absorbing capacity	High glass fibre content in deep ribs, low pressure forces
Application	Bodyparts in the automotive industry, oil pans	Bumper carrier, underfloor systems	Dashboard carriers, technical front-ends, bumper carrier, underfloor systems

Thermoforming

Thermoforming is a manufacturing process where thermoplastic sheets are heated up to temperatures that allow forming of the material to a certain shape. The semi-finished material consists of either non-reinforced polymers, particle, short or long fibre-reinforced polymers or of continuous fibre-reinforced thermoplastics.

The semi-finished sheets are fixed above a mould and heated by a contactless heating technology, e.g. infrared heating. Several thermoforming variants are available for forming the sheets. The sheets can be shaped using a male mould (die forming), air pressure (diaphragm forming), a rubber stamp, or a vacuum bag (vacuum forming) (Neitzel and Mitschang, 2004). The press opens after the solidification of the material. After the forming process trimming of the part is usually necessary although recent studies showed that net-shape produced continuous fibre-reinforced components are possible. Table 8.4 gives an overview for the most commonly used thermoforming variants.

8.3.5   Injection moulding

The injection moulding process is the most important process for the production of moulded parts made of thermoplastics, thermosets or elastomers not only in the automotive industry but in all industries using polymers (Michaeli *et al.*, 2001). Although injection mould tooling is expensive, the technology

Table 8.4 Comparison of thermoforming variants

	Die forming		Diaphragm forming
	Metal stamp	Silicone stamp	
Pros	Easy forming of small radii and edges High component reproducibility Extremely short cycle times High durability of the tool	Homogeneous pressure distribution Slight component undercuts possible Extremely short cycle times Less costs compared to metal stamp process Variable tool cavities	Prepregs processible Slight component undercuts possible Slight matrix oxidation Minor buckle formations Only one mould half Different laminate thicknesses
Cons	Holding-down device necessary Exact compliance of the tool cavity No pressure built-up perpendicular to pressing direction High tool costs	Holding-down device necessary Decreased component complexity compared to metal stamp process Special silicone stamp necessary	Long cycle times High maintenance effort and costs for diaphragm Residual stress during cooling down

is favoured due to the fully automated process together with the high part count and the excellent reproducibility of moulded components. The weight of producible parts ranges from a few milligrams up to 100 kilograms. The process in general is as follows: the granular material is plasticised typically using an extruder and moved through the machine using a special crew configuration. The melt is gathered at the front of the nozzle and is then forced into the tooling. After the mould is filled, a hold pressure is applied until the melt is solidified to prevent shrinkage. As soon as the part is solid the mould opens and the part is ejected. Figure 8.2 shows the process cycle. One can see that the total cycle time is determined, in particular, by the cooling time. Figure 8.3 shows the sketch of an injection moulding machine. The main parts are the clamping unit, the mould, the injection unit and the control system.

When working with reinforced thermoplastics fibre lengths in the range of 10–12 millimetres are possible, but due to the induced directional forces and shear forces applied during the plastification process a reduction in fibre length occurs. Additionally a strong fibre orientation exists in reinforced injection moulded parts which is related to high shear and directional flows in the mould depending on the geometric properties of the fibres and the viscoelastic properties of the matrix (Pötsch and Michaeli, 2007).

8.4   Fields of application for fibre-reinforced polymer composites (FRPCs)

Since the very beginning of the automotive industry polymers have played their part. The first cars often were made of a wooden chassis, one of nature's

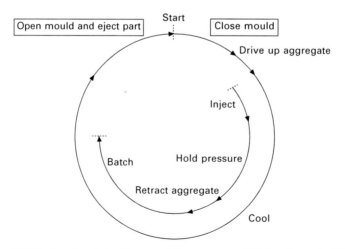

8.2 Cycle of an injection moulding process (Michaeli *et al.*, 2001).

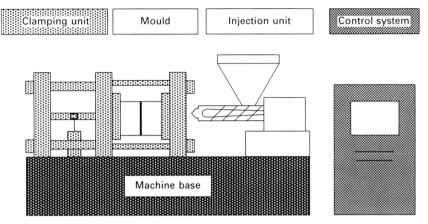

| Clamping unit | Mould | Injection unit | Control system |

Machine base

8.3 Schematic drawing of an injection moulding unit (Michaeli *et al.*, 2001).

most successful fibre-reinforced composites. Over a period of time while the requirements of cars changed, the applications for polymers changed as well. Structural parts were mostly replaced by metal and plastics diversified into other applications such as electrical, interior, exterior as well as 'under the hood' applications. In modern cars the amount of plastics used in each automobile varies from around 14% to over 20%. Figure 8.4 shows typical weight distribution for a range of cars produced by Audi.

Figure 8.5 shows the distribution of applications divided into four groups: electrical/light, 'under the hood', exterior and interior. As can be seen the majority of plastics is used for interior applications. Typical interior applications are seats, instrument panels, steering wheels and interior linings. Exterior plastic application examples are bumpers, roof modules, exterior mirror housings and car body panels whereas applications 'under the hood' describe parts like engine covers, air ducts, tubes, water tanks and hydraulic hoses. Electrical and light applications are found in headlamp front lenses, headlamp casings, electrical wire isolators and fastening clips. Altogether in an average car around 2000 parts are made of polymer materials of which about 1000 are used for joining elements (VKE, 1999).

8.4.1   Interior applications

Interior applications are dominated by plastic components. The reason for the success of polymeric materials is not only because of their lightweight character but is also explained by their widely adjustable visual appearance, the feel as well as acoustic behaviour and odour. All terms indicated are key criteria for quality perception and therefore important to establish a certain character of the brand. An example combining all the criteria is the instrument

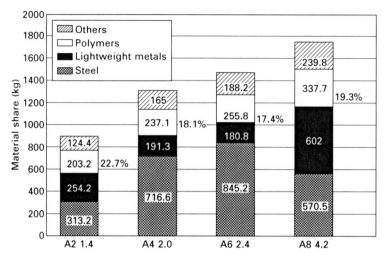

8.4 Weight distribution for various models produced by AUDI (Steuer, 2007).

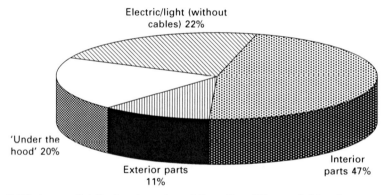

8.5 Polymer distribution in automobiles after different fields of application (VKE, 1999).

panel. While wood or leather might only be employed in upper class cars, many films and injection moulded surfaces try to imitate the texture and grain based on leather. To obtain certain surface finishes and feel, various technologies are currently in use ranging from simple injection moulding to more complex moulded skin technologies (see Fig. 8.6).

The process of manufacturing instrument panels works with either closed moulds or open moulds, each technology having its pros and cons. While closed moulds allow very precise wall thicknesses even in difficult regions, higher investment costs and a reduced design freedom are disadvantages compared to open mould technologies. On the material side both thermoplastic and reactive systems are utilised. PVC, thermoplastic polyurethanes (TPU), olefinic

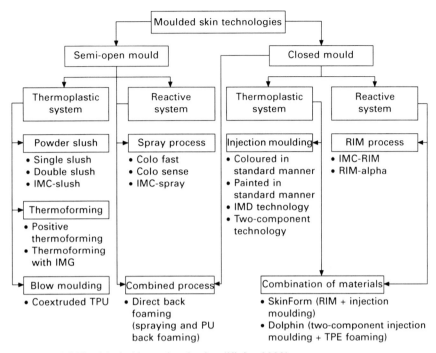

8.6 Moulded skin technologies (Klein, 2009).

elastomers (TPO) and reactive polyurethanes (PU) are mostly in use (Klein, 2009). In lower end passenger cars classic injection moulded polypropylene (PP) parts gain more and more importance. Since material suppliers have been able to adapt the required properties of PP some engineering plastics and even metals can be displaced. This led to many visible applications for PP-based components. Figure 8.7 gives an example of PP success for the Volkswagen Golf.

Another example of innovative plastic components can be found in seat technology. New materials like super-absorbing non-wovens are being integrated underneath the textile seat covering leading to reduced weight and costs while having an excellent air-conditioning effect compared with the mechanical ventilation systems currently used for air-conditioned seats (Petry and Rau, 2010). Even highly stressed parts can be produced using plastics as the Opel Insignia OPC driver's seat shows. It was developed in a joint development by Opel, Recaro and BASF and uses a glass-fibre filled, high impact modified polyamide to form the seat shell and the seat back by injection moulding. It combines a high degree of design freedom and comfort as well as sportive and stiff properties. To fulfil the legal as well as the OEM-specific crash requirements an intelligent geometry design combined with an appropriate material selection (a combination of

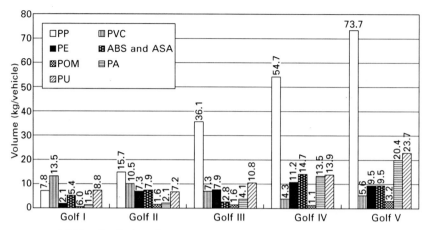

8.7 Plastics in interior applications in the Volkswagen Golf (Klein, 2009).

PA and energy-absorbing expanded polypropylene (EPP) foam) was used. A few years ago a development like this would not have been possible and metal would have been the material of choice, but the improvement of the computer-aided engineering (CAE) process and material models combined with material developments has enabled a mass volume production solution in polymeric material to be realised. The next step, clearly, would be to integrate continuous-filament reinforcement to locally increase stiffness. The fitting process technology might be an automated tape-placement process, which allows controlled tape-placement in order to obtain optimised mechanical properties in stress direction (Petry and Rau, 2010).

When less expensive parts are needed Twintex®(glass/PP commingled) is marketed as an excellent reinforcement. In a sandwich construction with Twintex® skins and polypropylene honeycomb cores it offers a good stiffness to weight ratio and is used, for example, in the Nissan (UK) Primera Estate as the trunk floor (Gardiner, 2006). Another well-known but still noticeable application is the backseat structure of the BMW M3 CLS which is made by thermoforming a sandwich of Twintex® and a poly(arylene ether sulfone) (PES) foam core. It shows the performance capability when novel materials and modern processes are combined with new applications (Doriat, 2005).

Finally it has to be mentioned that there are multiple applications where only plastics fulfil the requirements. For example, airbags, safety belts, clips of every kind, lateral covers and so on.

8.4.2   Chassis and exterior applications

Most of today's chassis applications are dominated by metallic components. Nevertheless there is a great weight saving potential when standard materials are substituted by polymeric solutions. In 2004 Bartus and Vaidya designed, modelled and manufactured a mass transit floor structure made of glass fibre (GF)/PP that reduced weight by 40 % compared to a standard steel/plywood structure. The chosen production process is carried out with a double belt press which is designed for higher output. Additionally the high cost of maintenance, corrosion and deterioration issues is eliminated when using glass/polypropylene thermoplastic composites (Bartus and Vaidya, 2004).

When it comes to exterior applications often surface quality becomes an issue. Lower and vertical parts like bumpers and side skirts are well established as plastic components. They are usually injection moulded and mainly consist of PP, PC or blend systems of those polymers. More critical are horizontal components like bonnets or roofs, where class-A surface qualities have to be guaranteed. A roof made of polymer composites, like the roof of the BMW M3, does not only reduce the car's weight but also influences the car's dynamic as the centre of mass is lowered.

MG Rover developed a bonnet that consists of unidirectional glass fibres in a polypropylene matrix that meets class-A surface requirements. The combination of unidirectional upper layers and a core of non-woven GMT proved both, economically and technically viable. The selected thermoforming process with its low tooling costs, which was used to manufacture the bonnets, is economical between low to medium quantities. Unfortunately, the thermoplastic hood never went in service due to the insolvency of MG Rover in 2005 (Gardiner, 2006). For thermoset systems there is a composite bonnet on the market used for the Aston Martin DBS. It is produced by Gurit and uses their Sprint CBS®, which is a thermoset system manufactured using a resin infusion technology. To obtain class-A quality a surface film is used on the top layer. With the reinforcement being either a glass/carbon mix or pure carbon fibres Gurit claims a weight saving from 56% to 70% compared to steel bonnets (Griffiths, 2010).

For established processes like SMC new advances were made using carbon fibres as reinforcement material. An application using the so called 'Advanced SMC' developed by Menozolit Compounds International GmbH is the three-piece scuttle panel on Daimler's Mercedes SLR Silver Arrow sports car (McConnell, 2008). The use of heavy tow carbon fibres allows a weight reduction of up to 60% compared to metal.

Material and process developments over the last few decades have allowed headlights, other lamps and windows around the car to be made of plastics. The use of plastics and therefore injection moulding processes offered an improved design freedom which is now driven even further with the introduction of high performance LEDs. The materials, however, have to withstand high temperatures when, for example, used in headlamps. For large headlamps glass fibre-reinforced polypropylene is usually sufficient to meet the heat resistance requirements. For other headlamp housings with more complex requirements higher quality thermoplastics, such as polybutylene terephthalate (PBT) or PES, are needed. The lighting unit is, of course, also subject to extremely high temperatures. The reflectors are therefore made of either sheet metal or metallised injection moulded thermoset (BMC) or amorphous high-temperature thermoplastics such as high temperature resistant polycarbonate (PC-HT), polyetherimide (PEI), PES and poly(phenylene sulfone (PSU)). BMC and thermoplastics compete with each other, with thermosets being much less expensive, while HT thermoplastics offer advantages for the manufacture and design of reflectors (shorter cycle times, excellent surface qualities, significant weight savings, complex geometries with functional elements and recycling capabilities) (Queisser *et al.*, 2002). A recent development in this area is shown in Fig. 8.8.

An upcoming area for automotive applications is glazing. Components based on polycarbonate have been in service for many years at Daimler/ Chrysler for fixed side windows and roof applications. The actual roof module of the Smart Forfouris is also manufactured using polycarbonate as a glazing material (Lehner and Aengenheyster, 2007). The outer regions of the roof module have coloured polycarbonate integrated which is produced using a combination of injection moulding and injection compression moulding. A combination of these two processes is not only complex to design but also requires a lot of technical know-how and is difficult to simulate as simulation programs handle injection moulding and compression moulding separately. Novel CAE methods are now able to link the two processes (Shuler, 2010). The first polycarbonate windows were rather small (up to 0.5 m²) and could be produced using conventional injection moulding. However, with larger parts (up to 1.2 m²) the aforementioned combination has to be used, resulting in lower injection pressures and reduced inner tensions. As large windows

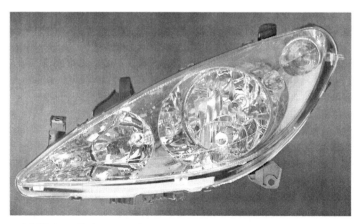

8.8 PBT headlight bezel from the Peugeot 307 (Bluhm *et al.*, 2008).

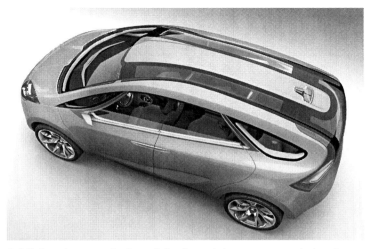

8.9 Polycarbonate glazing of the 'i-mode' concept car. A development between Hyundai and Bayer MaterialScience AG (Bayer AG).

often have darkened regions, this darkening is manufactured by the two-component moulding technique (Shuler, 2010).

However, abrasion and scratch resistance and therefore hardness of polycarbonate are still a matter of research. The application of PC glazing for front windows is not possible and mineral glass remains the material of choice. An often used method to improve the durability is a coating applied by plasma CVD treatment with considerably higher abrasion and scratch resistance (Lehner and Aengenheyster, 2007). But also hard coatings on the basis of melamine and silicate structures applied by sol-gel process are under development (Shin *et al.*, 2008).

8.4.3   Body applications

Within the last decade body applications have been dominated by metal as a construction material. The passenger cabin, especially, is built of various sorts of steel ranging from low strength to ultra-high strength steels with a tensile strength up to 1900 MPa. Only some extreme sports cars like the Porsche Carerra GT or Mercedes SLS come with a carbon fibre-reinforced polymer composite (CFRPC) monocoque. Nevertheless, some applications can be found that take advantage of continuous fibre-reinforced composites.

The German company Jacob Plastics GmbH, for instance, manufactures bumper beams made of Tepex® polyamide 6 plates for the BMW M3 by thermoforming. The thermoformed parts are assembled afterwards using induction welding, as a major advantage of thermoplastics is their weldability (Doriat, 2005).

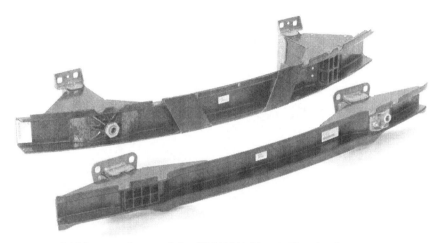

8.10 Bumper beam of the BMW M3 (Jacob Plastics GmbH).

A more recent development is the use of thermoplastic continuous fibre-reinforced composites (organic sheets) in the hybrid frontend module of the new Audi A8. The lower boom uses Tepex® plates with a PA6 matrix which is thermoformed into u-shaped profiles and then placed into the injection mould, where the front end is produced. Compared to an aluminium hybrid front end the use of organic sheets saves about 10% of the component's weight (Lanxess AG, 2010).

Within the engine compartment various pipes, housings and covers are manufactured from mainly engineering plastics. A more sophisticated application, however, is the engine mounts in the Opel Insignia. The injection moulded PA6 parts allow weight savings of 50% compared to the previous aluminium components (Rau, 2009). As material suppliers were able to increase the service temperatures of polyamides to 200 °C, applications like oil pans became possible. The Mercedes Benz Actros truck uses such an oil pan and achieved a weight saving of about 1.1 kg per part compared to aluminium. As the requirements for oil pans are not just high temperatures but also high impact resistance (rock impact and gravel on the road) and being leak-free, the use of plastics in this area is an interesting development and a promising challenge. A look into applications for high performance polymers also shows new developments as well. Mercedes, for example, introduced an air intake module for their V6 and V8 engines that is produced from a combination of polyphenylene sulfide (PPS) and PA46. And although the material price is higher compared to the substituted metal the total manufacturing costs could be reduced by 30%. This is possible by a so called assembly injection moulding process that permits simultaneous injection moulding and fully automated assembly of eight components. Manual finishing or assembly is no longer required (Reuschel and Hess, 2009).

8.5   Further challenges for composites in the automotive industry

The demand for lighter, safer and more efficient cars will secure and enhance the role of plastics in the automotive industry in the future as they still offer an excellent weight to performance ratio. This development is leveraged by the fact that most OEMs diversify their lines of products leading to lower quantities for each series. As production costs are important factors for the economic efficiency fewer quantities make composite materials (including their process technologies) more interesting.

Although classical power units will remain standard for the next few decades an opportunity offers the change from fossil driven engines to electric vehicles. In 2010 China sold about 18 million electric cars while the German government plans to have about one million electric cars on the road by 2020 (Bohrer, 2011). Potential applications for plastics in such cars might be component capsules, insulations, fixtures as well as control units and circuits (Anon., 2009).

As research is currently being done by thermoset material suppliers to decrease curing times, future automotive applications will make more extensive use from out-of-autoclave techniques like high pressure RTM (HP-RTM). This and variations of this technique will allow the cost-efficient manufacturing of several tens of thousand parts per year.

The combination of individual processes like thermoforming and injection moulding, a development that was recently seen at the 2010 K trade fair, will increase in the future. It offers synergetic effects leading to faster, cheaper and higher performance components which in addition allow intelligent integration of functions.

Another challenge for future automobiles might be the increased use of bioplastics such as polylactic acid (PLA) and bio-based polyamides and polyesters. Mainly Japanese companies such as Honda, Mazda and Toyota already use bioplastics for automotive components and new automotive applications are being developed in the field of textile seat covers, upholstery or surfaces, as PLA for instance offers good breathability, gloss and good tactile properties. Additionally natural fibre-reinforced composites will remain at their actual level of use, while having a reasonable low price and offer high performance at low weight (Kab, 2009).

For continuous fibre-reinforced composites a key factor for their success is the ability to model the material behaviour. This includes static as well as dynamic loads including crash simulations. And although there have been a lot of improvements in the past few years material models still have to be further developed to properly represent the material behaviour. This development will ensure composites have a bright and diversified future.

8.6   References

Anon. (2009), Potential for Innovation in the Cars of Tomorrow, *Kunststoffe International*, 3, 44–46.

AVK (2004), *AVK-TV-Handbuch 2004*, Frankfurt am Main, Federation of Reinforced Plastics e.V.

Bartus S.D. and Vaidya U.K. (2004), Design and Manufacture of Woven Reinforced Glass/Polypropylene Composites for Mass Transit Floor Structure, *Journal of Composite Materials*, 38, 1949–1972.

Bhattacharyya D. and Fakirov S. (2007), *Handbook of Engineering Biopolymers, Blends and Composites*, Munich, Carl Hanser Verlag.

Bluhm R., Eipper A. and Fitta I. (2008), The Right Symbiosis – High-performance Plastics in Headlights, *Kunststoffe International*, 3, 35–37.

Bohrer W.P. (2011), *Steht das Elektroauto am Anfang eines Siegeszuges*. Available from: www.eu-select.com/news_inhalt.php?id=2011_03_28 (Accessed 08 April 2011).

Chou T-W. and Ko F. K. (1989), *Textile Structural Composites*, Vol. 3, New York, Elsevier.

Davis B., Gramann P., Osswald T. A., Rios A. (2003), *Compression Molding*, Munich, Carl Hanser Verlag.

Doriat, C. (2005), Additional Knowledge Instead of Competition, *Kunststoffe Plasteurope*, 3, 1–6.

Gardiner G. (2006), Thermoformable Composite Panels: Part II, *Composites Technology*, 3.

Griffiths B. (2010), Gutir CBS for the Aston Martin DBS, *Composites Technology*, 1.

Kab H. (2009), A Logical Development – Bioplastics, *Kunststoffe International*, 8, 6–11.

Karus M., Ortmann S., Gahle C. and Pendarovski C. (2006), *Use of Natural Fibres in Composites for the German Automotive Production from 1999 till 2005*, Hürth, Nova Institut.

Klein, B (2009), Trends in Automobile Interiors, *Kunststoffe International*, 3, 56–59.

Lanxess A.G. (2010), *LANXESS: Frontend in Hybridtechnikmit Organoblech, Leverkusen*. Available from: http://plasticker.de/news/shownews.php?nr=10551 (Accessed 11.04.2011).

Lehner E. and Aengenheyster G. (2007), 'Polycarbonate Automotive Glazing: Automotive Industry Requirements and Solutions', in Stauber R. and Vollrath L., *Plastics in Automotive Engineering*, Munich, Carl Hanser Verlag, 335–345.

McConnell V.P. (2008), *New Recipes for SMC Innovation*. Available from: http://www.reinforcedplastics.com/view/1742/new-recipes-for-smc-innovation/ (Accessed 16.04.2011).

Michaeli W., Greif H., Kretzschmar G. and Ehrig F. (2001), Training in *Injection Molding*, Munich, Carl Hanser Verlag.

Neitzel M. and Mitschang P. (2004), *Handbuch Verbundwerkstoffe*, Munich, Carl Hanser Verlag.

Petry M. and Rau W. (2010), Are You Sitting Comfortably? – Lightweight Seats, *Kunststoffe International*, 3, 27–30.

Pötsch G. and Michaeli W. (2007), *Injection Molding*, Munich, Carl Hanser Verlag.

Queisser J., Geprags M. and Bluhm R. (2002), Trends in Automotive Headlamps, *Kunststoffe Plasteurope*, 92, 32–34.

Rau W. (2009), Lightweight Design Using Customized Engine Concepts, *Kunststoffe International*, 3, 47–50.

Reuschel G. and Hess J. (2009), Plastics in the Fast Lane – Metal Replacement, *Kunststoffe International*, 6, 46–48.

Shin Y.J., Oh M.H., Yoon Y.S. and Shin J.S. (2008), Hard Coatings on Polycarbonate Plate by Sol-Gel Reactions of Melamine Derivative, PHEMA and Silicates, *Polymer Engineering and Science*, 48, 1289–1295.

Shuler S. (2010), Auf Lange Sicht – Polycarbonat, *Kunststoffe*, 11, 96–99.

Steuer U. (2007), 'Increased Use of Plastics in Body Applications: Opportunities and Risks', in Stauber R. and Vollrath L., *Plastics in Automotive Engineering*, Munich, Carl Hanser Verlag, 7–18.

VKE Verband Kunststofferzeugende Industrie e.V. (1999), Kunststoff im Automobil – Einsatz und Verwertung, Frankfurt, selfpublishing.

9

Advanced automotive body structures and closures

P. URBAN and R. WOHLECKER, Forschungsgesellschaft
Kraftfahrwesen mbH Aachen, Germany

Abstract: This chapter begins with an overview of current design concepts for car body structures and explains the dependencies between those design concepts, production volumes and material application. It also introduces some important manufacturing technologies, semi-finished products as well as specific design concepts for closures. Based on a review of key factors bringing about change, three major trends in material usage for car bodies are identified and discussed. The chapter also gives examples of latest technologies demonstrated in two major research and development projects, which present an outlook on possible future material applications in car bodies.

Key words: design concepts for body structures and closure panels, costs of lightweight design, crashworthiness, high-strength vs low-density materials, multi-material solutions on different levels.

9.1   Current technology, applications and vehicles

During the development process of a series vehicle body a multitude of requirements has to be considered with regard to stiffness, energy-absorbing capability and structural integrity in a crash, NVH behaviour, durability, surface quality, corrosion resistance, production costs and recyclability amongst others. Apart from the design of the body and its components the material selection plays a key role in the fulfilment of these requirements.

Up to now the steel shell design has prevailed for most of the automotive bodies produced worldwide. This design concept is characterised by the application of stamped steel sheets which are joined mainly by resistance spot welding (Fig. 9.1). For several years, more and more laser welding and adhesive bonding have been introduced as additional joining techniques. Partially up to four steel sheets are joined and form profiles like the side sill or shear panels like the firewall. Corrosion protection is guaranteed by electro galvanised coating or by hot-dip galvanising. An outer skin made of stamped steel sheets is also very suitable for providing high-quality (class-A) surfaces.

Typical steel bodies for vehicles of the C-segment (e.g. Volkswagen Golf) weigh about 260 kg without closures. This value represents about 20% of the

230

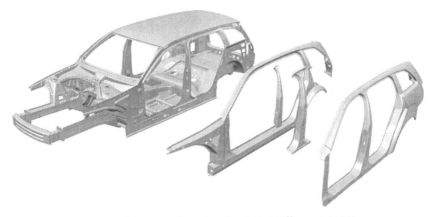

9.1 Steel shell design of vehicle bodies (Hoffmann 2009).

typical kerb weight in this vehicle segment. The closures add about another 100 kg to the body weight. In total, about 350 single sheets are joined to form a typical steel body.

In a very few vehicles the body shell is designed with aluminium instead of steel. The design principle is very similar, just the used profiles are designed with a larger cross section area. A typical representative of that design is the Jaguar XJ. Furthermore, the body of the Honda NSX sports car is manufactured from aluminium.

For small production volumes, as seen in sports cars and luxury vehicles, often different body design concepts are applied. These are mainly the tube frame design and the aluminium space-frame. Contrary to the steel shell design, in the tube frame design, the outer panels do not fulfil structural functions, so other materials are frequently used, in particular, various types of plastics (Fig. 9.2). Furthermore, in small amounts, ladder frame structures are used mainly for small commercial vehicles like pick-up trucks or for some off-road vehicles.

The aluminium space-frame design was introduced in 1994 by Audi in the first A8. It was produced with a volume of about 15,000 per year and its design is typical for the application of aluminium in the body structure. The third generation of this aluminium body design is running on the streets today. Apart from the main parts, extruded aluminium profiles, aluminium sheets and castings are used in this latest version; only the B-pillars are made of steel (Fig. 9.3). A substantial weight reduction compared to conventional steel designs of similar vehicles is possible with aluminium space-frames. The larger space needed in the profiles has to be considered. The aluminium space-frame design is also applied, for example, in the Audi A2 and the Rolls-Royce Phantom as well as in other small series production vehicles. The joining of aluminium components is done by riveting or clinching,

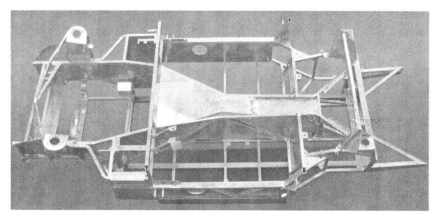

9.2 Steel tube frame design of vehicle bodies (Courtesy Wiesmann GmbH, Dülmen, Germany).

9.3 Aluminium space-frame design of vehicle bodies (Fidorra, 2010).

but welding is also possible. Typically the aluminium space-frame design is used for a production volume up to about 50,000 cars per year. This is mainly due to the better profit ratio of the cost-efficient extrusion tools at low production volumes in combination with the relatively high material price of aluminium, both in comparison to a steel shell design.

Another body concept that is primarily used in sports cars is the monocoque structure. The central structure is based on a closed body shell, that encloses the passengers by a complex structure. In the Porsche Carrera GT, for example, it consists of carbon-fibre-reinforced plastic (CFRP). The front- and rear-crash management systems of this vehicle are manufactured out of aluminium and stainless steel. In the event of a crash they are able to absorb most of the kinetic energy. In addition, they are easier to repair than CFRP structures (Eckstein, 2010). Crash management systems based on fibre-reinforced plastics (FRP) offer the advantage of even higher specific energy absorption capability under uniaxial loading. But, their application requires solutions to avoid the critical loss of structural integrity, which typically occurs during the crushing process of energy-absorption elements made of FRP. In addition, more sophisticated numerical crash simulation methods which enable reliable predictive analyses of the failure and post-failure behaviour of FRP structures are urgently needed.

Another opportunity of manufacturing monocoques is shown by the Wiesmann GT MF5. The carrying structure is a bonded and clinched aluminium monocoque (Fig. 9.4). The parts are produced without additional tools in a small series production by modern dataset-controlled laser, edging and milling technologies.

Up to now, the steel shell design has been further developed constantly in terms of applied steel grades, semi-finished products and manufacturing technologies as well as in terms of joining technologies. One of these further developments is Tailor Welded Blanking technology. This technology allows combining different steel grades, different sheet thicknesses and different surface qualities by laser welding two or more blanks to form a 'tailored' semi-finished product before stamping. Thus, the suitable material can be applied at the location where it is necessary without the need to join different stampings in a subsequent process step. Tailor Welded Blanking technology is applied, for instance, at the B-pillar of the current Audi A5 (Fig. 9.5). This B-pillar consists of a manganese-boron steel (1500 MPa tensile strength) in the upper area, while the lower area is made of a micro-alloyed steel (

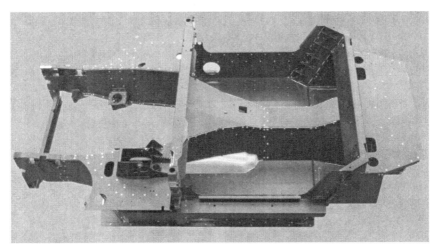

9.4 Monocoque aluminium design of vehicle bodies (Courtesy Wiesmann GmbH, Dülmen, Germany).

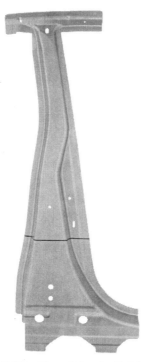

9.5 B-pillar of the Audi A5 with Tailor Welded Blanking technology
(Adam, 2008).

500 MPa tensile strength) (Adam, 2008). In a side impact, this solution
combines controlled deformability and energy-absorbing capability in the
lower area with high rigidity for minimum intrusions in the upper area.

Tailored rolled blanking technology can be an alternative to tailored welded
blanks. In this case, the semi-finished blank is manufactured by a flexible
rolling process with a variable roll gap, so that no welding step is necessary.
Thus, sheet components with continuous transitions of thicknesses can be
realised. However, a sheet thickness change is just possible in the rolling
direction. And, the usage of different steel grades is very limited. Different
strength values can be realised only in combination with tailored tempering,
which means that merely a part of the component is tempered in order to
increase its strength locally.

Furthermore, various alternative technologies to stamping like rollforming
and hydroforming are used in single components of automotive body structures
today. With rollforming it is possible to produce endless steel profiles with
nearly any cross section. In the process a steel sheet runs through a set of
rolls which are arranged in a special configuration to create the required
cross section. The resulting profile is then cut at the required length and can
be used, for example, as a side sill. Due to low formability requirements,

very high-strength steel grades can also be processed by roll forming. Apart from that, the tooling costs are very low.

With hydroforming the shape of a steel tube can be varied in a very flexible way and the geometry of the component can be fine-tuned (calibration). The semi-finished material, mostly a straight or bent tube section, is put into a two-piece die. The cavity of the die corresponds to the contour of the finished component. The die is closed and kept shut by a hydraulic press. Hydraulic punches seal the tube ends and fill the tube with fluid. In the actual forming process, the tube is inflated by the pressure build-up. Simultaneously the seal punches compress the tube ends. Thus, the material flows into the forming zone. In the last phase of the process, the interior pressure is increased, so that the workpiece gets its final shape, which corresponds to the contour of the die accurately (calibration step) (Eckstein, 2010).

Other body materials than steel, in particular aluminium and plastics, are found mainly in low- and extra-low-volume production today. This is partly due to the relatively high tooling costs of steel stampings, which in high-volume production are easily compensated by lower material costs. Moreover, the optimisation of vehicle properties by lightweight design has generally higher priority in the market segment of premium brands than in the more cost-driven high-volume market. For good vehicle handling characteristics, a balanced distribution of axle loads and a low centre of gravity are important. This is the reason why lightweight design usually has highest importance in the front section and in the roof area of front-engine cars.

The BMW 6 Series represents a vehicle that contains both aluminium and different types of plastics in these areas. A thermoplastic polymer blend is used in the front fender reducing its weight by about 2 kg per fender compared with a corresponding steel part. The deck lid of this vehicle is manufactured of thermoset sheet moulding compound (SMC), which guarantees the necessary styling freedom, integrates the antennas and needs no corrosion protection. A weight reduction compared with steel of about 2.5 kg is achieved, although lightweight design is not the top priority for this component (Gruenn, 2004). The opposite is true for the front structure of the vehicle, which is entirely made of aluminium including the main components strut towers and longitudinal beams, while the remaining body structure is made of steel. Also the doors, the bonnet, the strut tower beam and the front bumper are aluminium parts. The manufacturing techniques stamping, casting and extrusion are used to produce the various aluminium parts (Gruenn, 2004).

Vehicle body components manufactured in high-performance fibre-reinforced plastics are used mainly for vehicles with very low production volumes. Plastics with directed glass and carbon fibre reinforcements promise very good performance potential compared to steel and aluminium. However, time and cost-efficient manufacturing processes are not in place yet and the costs of raw material are still very high. Due to that there are just a few applications in series vehicles. Examples are the bonnet of the Audi R8 GT or the roof of the BMW 3 Series M3. Both are made of carbon fibre reinforced plastics. Manufacturers are undertaking significant work in this area and several new vehicle constructions are expected over the next few years, e.g. BMW MegaCity featuring the use of high performance fibre-reinforced composites at higher production volumes.

Typical vehicle closures (doors, bonnets, deck lids) are made of two steel shells. The inner shell is a stamped sheet which provides the structural performance of the closure. In doors it also carries the assembly parts and usually integrates the window frame. Some door concepts, mainly by Audi, however, use a separate window frame. The outer shell is a relatively flat sheet, which is responsible for the styling and the buckling behaviour of the closure.

Many modern cars, mainly in the upper vehicle class, feature aluminium doors. The inner shells of the Porsche Panamera doors each consist of an aluminium casting, for example. Another example of an innovative series application of aluminium in closures are the doors of the BMW 7 Series, which are completely made of aluminium (Fig. 9.6, top (E65) and bottom (F01)). Both the inner and the outer panel are manufactured by stamping aluminium sheet. A special feature of the inner panels is the application of a new sandwich aluminium material, consisting of three layers which are cast and rolled to an aluminium blank as a semi-finished product. This blank is

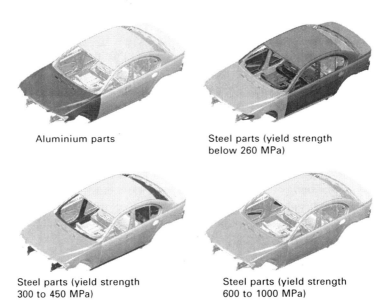

Aluminium parts

Steel parts (yield strength below 260 MPa)

Steel parts (yield strength 300 to 450 MPa)

Steel parts (yield strength 600 to 1000 MPa)

9.6 Material mix in the body of the BMW 7 Series (E65 versus F01) (Floeck and Pfestorf, 2008).

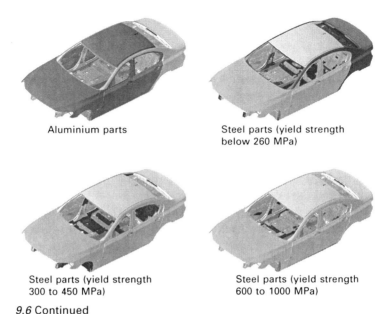

Aluminium parts

Steel parts (yield strength below 260 MPa)

Steel parts (yield strength 300 to 450 MPa)

Steel parts (yield strength 600 to 1000 MPa)

9.6 Continued

then stamped to form the finished component. The main advantage which the combination of different alloys in the covering layers and in the core of a single blank offers is the possibility to better balanced material properties like strength, formability and corrosion resistance. In the inner door panels of the BMW 7 Series, very small stamping radii amongst others are made feasible by a technology known as Novelis Fusion® (Bassi, 2009). BMW claim a weight of about 47 kg for the four door structures of the current BMW 7 Series F01. A weight reduction of about 28 kg per vehicle to a comparable steel design is achieved (Macha, 2008).

Since the introduction of the European Directive 2003/102/EC (EUR-Lex, 2003) in 2005 the topic of pedestrian protection has gained more importance. Meanwhile this directive transferred to regulation (EC) No 78/2009 (EUR-Lex, 2009), which became effective in 2009. It guarantees a defined minimum standard of pedestrian protection. This regulation influences the styling and the design of vehicle bonnets considerably. In order to fulfil this regulation structural improvements in the vehicle front are necessary. These improvements include the designing of larger radii, the avoidance of stiff areas at the impact locations and the provision of a sufficient distance between the bonnet and the hard points of the engine compartment in order to absorb the impact energy. In addition to purely passive solutions, deployable protection systems like pop-up bonnets are used in very few series vehicles. Furthermore, external pedestrian airbag systems are under investigation to face the more demanding requirements of the consumer protection organisation Euro NCAP.

9.2   Key factors driving change and improvements

Several key drivers and influencing factors are bringing about changes with regard to current material applications in vehicle body structures and closures. The development of new materials for the vehicle body structure and for components is going on continuously in competition between various suppliers, which is an important driver for technological improvements on its own. This section, however, will focus on non-technological factors which are influencing the selection of materials and the application of lightweight measures in body structures and closures. These are, amongst others, CO_2 emission regulations, the volatility of material prices, the global supply situation and the trend towards the electrification of vehicles.

The high costs of many lightweight technologies are still an important barrier for their series application. However, a light body structure contributes substantially to fuel consumption reduction and CO_2 savings, in particular if the potential of secondary weight reductions is taken into account. Intensive calculations for all vehicle sections show that a secondary weight saving of 30 kg is possible while a primary weight saving of 100 kg for a compact class vehicle is considered. Taking this weight reduction into account and considering further calculation loops another secondary weight reduction of 15 kg can be achieved (Goebbels, 2010).

The weight of a vehicle has a big influence on the fuel consumption and with it on the CO_2 emissions. The benefits of a weight reduction are basically shown by the power demand of a vehicle.

$$P = (m_v + m_l) \cdot g \cdot \cos(\alpha_{as}) \cdot f_R \cdot v + (e_i \cdot m_v + m_l) \cdot a \cdot v$$
$$+ (m_v + m_l) \cdot g \cdot \sin(\alpha_{as}) \cdot v + 0.5 \cdot \rho_{air} \cdot c_d \cdot A \cdot (v - v_w)^2$$

[9.1]

where m_v is the empty vehicle weight, m_l is the payload, g is the gravitational acceleration, α_{as} is the angle of ascent, f_R is the rolling-resistance coefficient, v is the vehicle speed, e_i is the inertia coefficient for the i-th gear, a is the longitudinal vehicle acceleration, ρ_{air} is the density of the ambient air, c_d is the drag coefficient, A is the cross-sectional area of the vehicle and v_w is the wind speed in driving direction.

Except for drag resistance, which can be reduced only by decreasing the drag coefficient and/or by the reduction of the cross-sectional area of the vehicle, every addend of the formula above contains the mass of the full vehicle (Eckstein, 2010). Intensive simulations help to quantify the influence of a vehicle weight reduction on fuel consumption. According to such simulations, a 10% mass reduction without powertrain re-sizing saves between 1.9% and 3.2% fuel in internal combustion engine vehicles when considering the new European driving cycle (NEDC) and hybrid technology development approaching efficient zero emission mobility (HYZEM). These values correspond with a fuel consumption reduction of about 0.12/100 km to 0.16 l/100 km per 100 kg weight reduction. A re-sizing of the powertrain in order to achieve the same acceleration as the reference vehicle would increase the fuel consumption reduction enormously (Wohlecker, 2007).

An important reason for an intensive lightweight design is the latest EU regulation 715/2007 (EUR-Lex, 2007) on CO_2 emissions. According to this regulation, the emissions of all new passenger cars in the EU shall be reduced to an average fleet emission of 130 g CO_2/km per vehicle until 2012 for each brand. Actions such as the more intensive use of bio fuels are considered to be equivalent to further reduction of 10 g CO_2/km. In the long-term, the limit is to be 95 g CO_2/km by 2020. There are penalties between €5 and €95 per gram CO_2 per vehicle from 2012 onwards, if these values are exceeded. These penalties have to be balanced with the extra costs of lightweight design and other CO_2 reduction methods. Similar regulations can be found in other regions of the world; especially in some areas of the USA there are very demanding regulations in place.

As a consequence, the break-even point where these lightweight technologies are affordable will probably shift to higher costs per kg weight saved in view of the penalties for exceeding the CO_2 limits and the expected long-term increase in the costs of resources like crude oil.

Currently the price for crude oil is around US$110 per barrel and has been so since the beginning of 2008 (Fig. 9.7). A high peak of about US$130 was seen in the middle of 2008. After that it came down to about US$40. The overall trend is rising, and since it is expected that crude oil resources will decrease, the price will probably continue to increase on a long-term basis. This perspective suggests that the efforts towards automotive lightweight design will intensify in the long run.

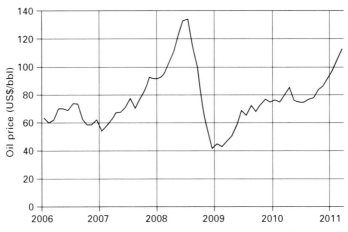

9.7 Trend of crude oil price in US dollars per barrel (INF, 2010).

Moreover, the main materials applied in automotive bodies, steel and aluminium, are subject to high price volatilities. The price depends on the specific material grade, but a general trend is clearly visible. The price of steel, shown in Fig. 9.8, came down from around €1100 per ton in 2008 to about €550 in 2009. It is currently around €800 per ton, with a rising trend. The price of aluminium is now close to the August 2008 level where about US$3000 per ton was reached (Fig. 9.9). The lowest level can be seen at the beginning of 2009 at around US$1300 per ton. Also with aluminium the price trend is rising. All in all, depending on the currency, the aluminium price is about two to four times higher than the steel price.

During the development process of a vehicle body which is to be produced and sold all over the world, the production possibilities and especially the

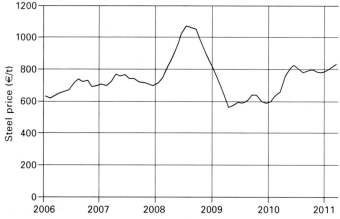

9.8 Trend of steel alloy prices in Euros per ton (LME, 2010).

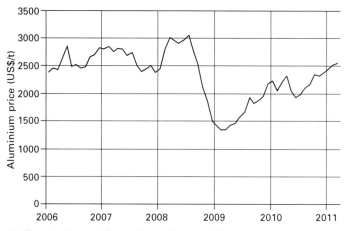

9.9 Trend of aluminium alloy prices in US dollars per ton (LME, 2010).

availability of the materials in the different regions has to be considered. For example, the availability of ultra high-strength steels is restricted in some Asian areas. Therefore, flexible design solutions in terms of the material application for certain body components need to be considered.

Furthermore, in future the expected market entry of electric vehicles will play an important role in terms of lightweight costs. The key factor in this context is the heavy and expensive battery. If the vehicle weight can be lowered by lightweight measures, smaller batteries will be sufficient for achieving the same operating range. Thus, the costs of lightweight measures have to be offset against the costs of saved battery capacity. In this consideration, it is important to take into account the energy recuperation capability of electric vehicles, which reduces the mass-dependence of the overall energy consumption of an electric vehicle compared to a conventional one.

A corresponding study by the Institut für Kraftfahrzeuge at RWTH Aachen University is based on the simulation of the energy consumption of a subcompact class electric vehicle (Hartmann *et al.* 2010). Three different driving cycles in combination with vehicle weight variations have been simulated. Based on that, the required battery size for a given operating range and the corresponding costs of each constellation have been determined. This leads to the possible battery cost savings per kilogram weight reduction shown in Fig. 9.10 (Hartmann *et al.* 2010). These results indicate that even expensive weight reduction measures can result in overall cost savings, depending on the assumed battery costs.

All in all, the trend towards an increasing importance of lightweight design is obvious from the key drivers and influencing factors mentioned above. However, the question remains how intensive this trend will be. Together

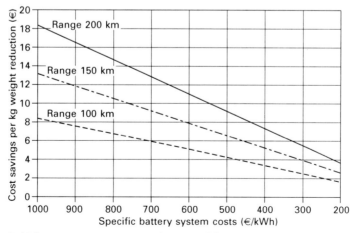

9.10 Battery system cost savings per kilogram electric vehicle weight reduction.

with the volatility of material prices and the global material supply situation, this question also drives an interest in more flexible solutions for the material usage in automotive bodies than those available today.

9.3   Trends in material usage

Basically three major trends in material usage for automotive bodies can be identified: the trend towards high-strength materials in the body structure, the trend towards low-density materials in closure panels and the trends towards multi-material solutions. All these trends result from the need to meet all performance requirements at minimum costs, while facing more and more ambitious weight targets.

The design of an automotive body structure and the corresponding material selection are today very much determined by crashworthiness requirements. Since the vehicle's kinetic energy has to be absorbed in a crush zone of limited size, for example when crashed into a rigid barrier, the affected structural members have to carry high impact loads without showing catastrophic collapse. Making this possible at low weight requires the application of materials with high strength-to-density ratios. For most OEMs, high-strength and ultra high-strength steels have become the materials of choice here in high volume production. This has been facilitated substantially by the steel industry's effort to offer a wider and wider range of sheet steel grades, in particular in the high-strength and ultra high-strength regimes with a minimum yield strength of 240 MPa respectively more than 550 MPa tensile strength (AISI, 2006). As an illustration, Fig. 9.11 shows the spectrum of fine sheet grades produced by a leading steel supplier. Most of the high-strength and ultra high-strength grades displayed in this diagram were introduced in the 1980s and 1990s (Weber, 2005), and the development of new, complementary

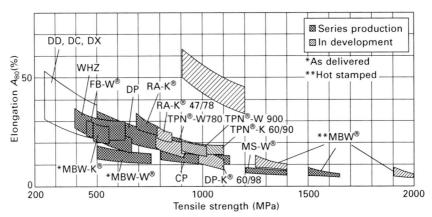

9.11 Failure strain versus tensile strength of sheet steel grades for automotive applications (Hoffmann, 2011).

grades with high- and ultra high-strength is continuing. Taking into account the typical development cycles of new vehicles, the automotive industry has been fast in adopting these new materials in its products, and the extent to which high-strength and ultra high-strength steels are used in automotive body structures continues to grow. As an example, Fig. 9.6 shows the material mix in the body of the BMW 7 series, F01 (bottom position) in comparison with the preceding model, E65 (top position). It is obvious from this figure, that, although the application of these materials is limited to structural parts, the percentage of high- and ultra high-strength steel usage in the complete vehicle body has almost doubled (from c. 29% to about 55%) within one model renewal cycle. Similar figures could be quoted for many other car models as well.

High strength, however, is not the only requirement which a material has to fulfil to be applicable in an automotive body by far. Another important requirement for sheet metal being processed in the press shop is sufficient formability to make sure that parts can be stamped without material failure. In addition, even when the part is formed, the finished component must still provide a component-specific level of residual formability or failure strain, so that it can contribute to the absorption of crash energy by controlled deformation. It is important to note here, that the basic function of a vehicle's crush zone is to turn kinetic energy into deformation work, which is the integral of force versus deformation distance. So, both a sufficient force level and a sufficient deformation distance are necessary to absorb a given amount of crash energy.

Figure 9.11 suggests a rough correlation between the strengths and the failure strains of the carbon steel grades displayed: the higher the strength, the lower the failure strain tends to be. In fact, mild steels usually allow for large plastic deformations, whereas ultra high-strength grades show reduced formability. To some extent, this can be compensated for by adapted component design reducing the need for high deformation levels. Another important approach is to induce changes to the microstructure of the steel during the manufacturing process of the component, so that good formability of the original blank can be combined with high-strength of the finished component. Depending on the steel grade, different mechanisms can take effect, for example work hardening due to plastic deformation during the manufacturing process, bake hardening induced by the heat introduction in the paint oven or press hardening in the hot forming process of manganese-boron steels. While all these hardening mechanisms are made use of in the series production of automotive body components today, there is currently a particularly strong growth in the number of press hardened components for the body structure. These components are heated up to a temperature of about 900 °C before the last step of the forming process and are subsequently quenched while still in the die. This hardening process results in parts with a martensitic microstructure, extremely high tensile strength and good geometric accuracy (Weber, 2007). While the first components produced by hot forming were bumper beams and side-impact beams, the process and the corresponding manganese-boron steels are now more and more making their way into the body-in-white, where components such as rockers, roof-pillars and other components with very high strength requirements have become typical applications of hot forming. The Volkswagen Group is one of the pioneers in massively introducing this process to large volume production, but other manufacturers are following. As an example of the significance that hot forming has gained in the production of car body structures, Fig. 9.12 gives an overview of the application of hot formed parts in the body of the VW Polo V. The steel industry's ability to continue the trend towards high-strength materials from the supply side while maintaining the cost advantages of steel over all other applicable materials will be key in sustaining the role of sheet steel as the most important vehicle body material in the long-term future. An important step in this direction is the on-going development of hot-forming steels with tensile strengths of up to 2000 MPa as well as cold-forming grades with strength levels of 1000–1500 MPa and failure strains of up to 50%.

Not all components of a vehicle body and not even all components of a body structure are determined by the requirements of crashworthiness alone. The suspension-strut domes, for example, have an important function with regard to the global stiffness of a car body, while their relevance for the vehicle's crash behaviour is relatively low. Therefore, the application of very high-strength materials would not allow for major weight savings here. Since all steel grades in use for the production of vehicle bodies today show approximately the same modulus of elasticity independent from their strength,

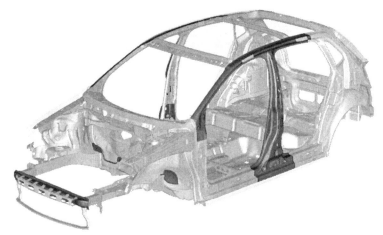

9.12 Hot-formed parts in the body of VW Polo V (Eichhorn, 2009).

the weight saving potential of high-strength steels is limited in applications which are rather critical with regard to stiffness than with regard to strength (EAA, 2010).

A similar consideration applies to closure panels. In fact, the wall thickness of these outer skin panels is to a large extent determined by the requirements for bending stiffness and local denting resistance. These properties depend more on a panel's wall thickness than on the stiffness or strength of the material applied. For this reason, there is a trend towards the application of materials with lower density than steel for these panels, even if the ratio of elasticity modulus to density, for example, may not be significantly better for these materials than it is for steel.

Aluminium sheet, in particular, is a low-density material with a growing share in the production of closure panels. Since its density is less than one third the density of steel, aluminium closure panels can be considerably thicker than corresponding steel panels to compensate to some extent for the poorer mechanical properties of aluminium and still allow substantial weight savings, if component strength is not the primary requirement. In fact, the growth rate which is expected for aluminium in car body sheet applications is three times higher than it is in structural applications (Hirsch, 2010). Typical aluminium sheet applications are both the outer and inner panels of side doors and bonnets, particularly in premium cars. Since there is no direct contact between unprotected aluminium and steel surfaces when assembling these components to the vehicle, potential problems with galvanic corrosion are avoided here. The same applies to luggage compartment doors and front fenders, of which the latter are often only screwed to the body structure for ease of replacement after 'fender bender' accidents. However, there are also first examples of aluminium panels which are fully integrated into a steel structure such as the roof panel of the BMW 7 series (F01) (Stauber, 2009).

Apart from aluminium, a variety of polymer materials is becoming more and more important in the production of 'hang-on parts', which are assembled to the vehicle body via detachable joints. Typical examples are front fenders, which cannot only be reduced in weight when made of polymers, but can also benefit from the elasticity of these materials in terms of damage tolerance. In addition, the injection moulding process of thermoplastic polymers facilitates complex component designs integrating numerous functions into one single component. The front fender module of the BMW X5 series (E70), for instance, makes full use of this potential for the integration of functions (Schenn, 2007). Another well-known example of the application of polymer body panels are the vehicles developed by the Daimler Group under the Smart brand and introduced into the market since 1998 (City-Coupé/Fortwo, Forfour and Roadster). However, the most important hang-on parts, which are made of polymer materials almost without exception today, are bumper fascias. In particular, polypropylene, polyurethane and polycarbonate are applied to produce these components (AISI, 2006). Due to the advantages of these polymers in terms of damage tolerance, corrosion resistance and the feasibility of very complex shapes, polymer bumper fascias cover a growing part of the whole vehicle's front and rear section.

All in all, the trends towards high-strength materials in the body structure and the trend towards low-density materials in closure panels directly result in the third major trend in material usage for automotive bodies: the trend towards multi-material solutions. This means that the advantages of different material classes are combined to deliver a spectrum of characteristics which would not be feasible with one single material class. The trend towards multi-material solutions can be observed on three different levels: on the full vehicle body level, the component level and on the material level.

On the full vehicle body level, there are hybrid vehicle bodies which consist basically of two major sub-structures made of dissimilar materials with a clear interface in between. Examples are the second generation Audi TT, which features an aluminium body with a rear structure made of steel, and the BMW 5 series introduced in 2003 (E60) with an all-aluminium front structure, which extends from the front bumper to the firewall and is integrated into a steel body. In both cases the hybrid body concept contributes both to weight reduction and to the realisation of a well-balanced distribution of axle loads in order to improve the handling behaviour of the vehicle. Multi-material solutions on the full vehicle body level can, however, also mean a more radical material mix, which implies that dissimilar materials are applied to various components of a vehicle body without forming major sub-assemblies. A good example of this approach is the body of the Porsche Panamera shown in Fig. 9.13. This four-seater sportscar not only features hang-on parts made of aluminium and, with even lower density, magnesium, but the steel body structure of the vehicle also integrates front longitudinal members, front strut towers and a tail panel made of aluminium, a plastic luggage compartment tub as well as a firewall cross member and several smaller parts made of stainless steel (Koehr, 2009).

In extra-low-volume production, multi-material solutions on the full vehicle body level have been in use for a long time as a consequence of the high importance of lightweight design in this market segment and the economic need to apply less investment cost intensive production technologies than stamping steel sheet. While the widespread application of plastic outer panels is likely to continue in extra-low-volume production, traditional steel tube frames as load-carrying structures are being replaced more and more by aluminium space-frames and by monocoque structures made of aluminium or CFRP. A good example of this trend are the body structures of the Wiesmann sportscars: while the Wiesmann Roadster is based on a tube frame structure, the coupé version, which was introduced about ten years later and features

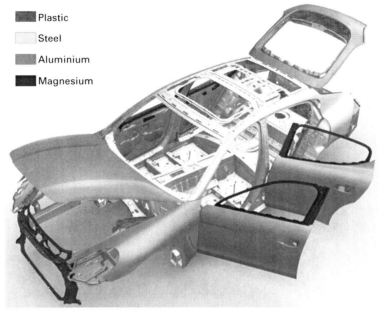

Plastic

Steel

Aluminium

Magnesium

9.13 Mix of materials in the body of the Porsche Panamera (Koehr, 2009).

almost the same visual appearance, comes with the aluminium monocoque shown in Fig. 9.4.

On the component level, hybrid solutions made of sheet steel and polymer materials are gaining more and more importance. Front-end carriers combining these materials have already been in high volume production since the late 1980s. There are several options when producing such components today; for example, by adhesively bonding stamped steel parts to an injection-moulded plastic carrier or by directly moulding the polymer into a steel sheet component. The functional principle, however, is always the same: The steel sheet provides the required strength to the component, while the polymer reinforces the sheet against buckling and integrates numerous attachment points for many components of the front end (bonnet latch and buffers, headlights, radiator, water hoses, fenders, front grille, bumper fascia, etc.) (Eckstein, 2009). Using such a metal-plastic front-end carrier as a core element of a front-end sub-assembly and fixing it to the vehicle after the complete powertrain has been installed enhances the accessibility of the engine compartment significantly during the assembly process and is a typical example of the successful implementation of modularisation strategies. In the meantime, the basic functional principle of metal-plastic front-end carriers has been applied also to other vehicle body parts. The Audi A6 introduced in 2004, for example, features a front roof cross member which is made as a steel sheet profile filled with a plastic reinforcement structure in order to avoid buckling.

Other types of hybrid components that are increasingly making their way into the body structure are composite inserts which can either have an acoustic or a reinforcing function. These inserts are placed in hollow sections of the structure in the body shop and consist of a metal or plastic carrier to which an expandable polymer is applied. The expansion of this polymer is initiated by the introduction of heat in the paint oven thus forming a firm connection between the insert and the surrounding profile. A particular advantage of these composite inserts is the possibility to include them into the structure at a relatively late point in the body development process.

On the material level, fibre-reinforced plastics constitute multi-material solutions with growing relevance in body engineering. A multitude of fibre-reinforced plastic components with no specific fibre orientation has already been in use for many years, mostly made of thermoset SMC, glass mat reinforced thermoplastics (GMT) or fibre-reinforced polyurethane. The use of directed fibre products such as fabrics is now opening the way for fibre-reinforced plastics into more and more semi-structural and structural components. An important enabler for the introduction of such high-performance composites into volume production are more cost-efficient processing technologies in combination with the growing relevance of lightweight design. Also the spectrum of applicable fibre materials is becoming broader. While glass fibre has been in use as a relatively inexpensive reinforcing material for a long time, the development of carbon fibres which are specifically designed for the requirements of the automotive industry is making carbon fibre an economically viable alternative not only in low-volume production. The first volume-production car with an all carbon fibre-reinforced passenger compartment will probably be BMW's Megacity Vehicle, which is expected to be launched in 2013 (Kranz, 2010). From a technical point of view, it is mainly the high stiffness-to-density ratio due to which carbon fibre stands out in comparison with glass fibre. Another interesting reinforcing material which has been introduced in series production recently is steel cord. The first application is the rear bumper beam of the Mercedes SLS AMG, which is a GMT component featuring an additional steel cord reinforcement for improved toughness (Pech, 2009).

Apart from fibre-reinforced plastics, steel sandwich blanks are also important examples of multi-material solutions on the material level itself. These blanks consist of two covering layers made of thin steel sheet and a plastic core layer in between. Even very thin core layers made of soft polymers can bring about major advantages with regard to acoustics by improving the damping characteristics of the blank. In contrast, thick core layers are a means to substantially increase the bending stiffness compared to a monolithic blank with the same weight thus constituting another interesting approach to lightweight design.

A challenge which is common to most multi-material solutions, in particular on the full vehicle level, is the challenge of reliably joining dissimilar materials even under thermal expansion and corrosion stress. Since, in many cases conventional welding techniques are no longer applicable, adhesive bonding and mechanical joining techniques such as riveting and clinching gain in importance. Further development of such joining concepts with a focus on cost-efficiency and particularly their representation in numerical simulation tools remain crucial for the application of multi-material solutions in high-volume production.

9.4   Latest technologies

Since several examples of advanced material usage in body structures and closure panels of production vehicles have already been given above, this section will focus on the latest technologies demonstrated in two research and development projects: SuperLIGHT-Car and InCar. These two projects present a good outlook on possible future material applications in automotive bodies.

SuperLIGHT-Car is an integrated research and development project co-funded under the Sixth EU Framework Programme for Research and Technological Development. The overall objective is to deliver the technologies and design concepts for an economically viable multi-material design of a compact-class vehicle body bringing about a weight reduction of 30% under the boundary conditions of high-volume production. The project was coordinated by Volkswagen Corporate Research and brought together seven European vehicle manufacturers plus more than 30 suppliers, engineering service providers and research institutes. The main achievements of the SuperLIGHT-Car project are as follows:

- Multi-material body-in-white concept offering a 35% weight reduction compared with a reference structure based on the VW Golf V.
- Design of the assembly line and joining sequence for this body-in-white concept.
- Full multi-material body-in-white prototype of this concept comprising many parts produced under industrial like conditions.
- Three alternative lightweight body-in-white concepts offering weight reductions by 20% to 38% at additional costs of €2.5 to €10 per kilogram saved.
- Multi-material assessment methodology to assist in the material selection process by classifying and matching the component requirements, material properties and manufacturing as well as joining aspects to identify the optimum material for a given part or sub-assembly.
- Materials and technologies catalogue and database with characterisation results of materials and joining techniques including crash toughened

adhesive, metal inert gas welding, laser (induction) welding and friction stir welding.

- Detailed cost models for multi-material concepts and life cycle assessment tool including a fuzzy logic recycling model, all linked to CAD software for quick A to B comparisons.

The final SuperLIGHT-Car concept is shown in Fig. 9.14. Approximately 53% of the total body-in-white mass consists of aluminium sheet, aluminium castings and extruded profiles. Hot and cold formed steel including tailor rolled and tailor welded blanks account for *c*. 36% of the total weight, while about 7% is made of magnesium castings and hot formed magnesium sheet. The remaining 4% consists of carbon and glass fibre-reinforced plastic parts. Some technological highlights include a tunnel with reinforcements made of press hardened tailor welded and tailor rolled steel blanks respectively, front longitudinal members made of aluminium tailor welded blanks, a centre roof crossbeam in carbon fibre-reinforced plastic pultrusion technology as well as a hot formed magnesium roof panel with front and rear roof crossbeams in the same technology. Other interesting solutions which do not only offer major weight savings, but also reduce the number of parts by the integration of functions are the high-pressure die cast magnesium strut towers and the rear floor in long-fibre reinforced thermoplastic (LFT) technology (Goede, 2009a, 2009b). With all these innovations in body engineering, the SuperLIGHT-Car project is expected to create an important momentum for the more widespread application of multi-material solutions in future high-volume cars.

InCar is an innovation project by the supplier ThyssenKrupp. It comprises a comprehensive set of innovations for the automotive body, chassis and

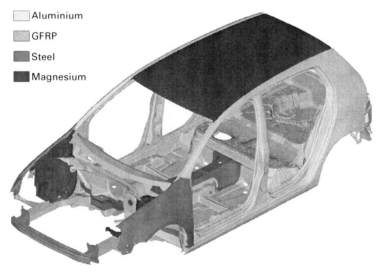

9.14 SuperLIGHT-Car final concept.

powertrain, each of which are adapted to different requirements in terms of weight reduction, cost-efficiency and functionality. The project follows the tradition of previous research and development activities by the steel industry such as the UltraLight Steel Auto Body (ULSAB), ArcelorMittal Body Concept (ABC), Atlas Spaceframe and NewSteelBody (NSB). It is unique, however, in the extent to which the various solutions have been analysed numerically and validated in terms of technical performance, costs, greenhouse gas emissions and manufacturing feasibility including the physical testing of many prototype parts.

Eight out of 17 sub-projects focus on various components and sub-assemblies of the vehicle body. For most of these applications, more than one solution has been developed, each of them showing a particular profile with regard to weight reduction, cost-efficiency and functionality and offering specific advantages compared with a manufacturer-neutral reference structure. This reference structure, which represents a state-of-the-art upper mid-size station wagon, has been developed within the project too.

A special feature of all these solutions is their modularity, which means that all concepts for a particular body application are interchangeable, so that the various components and sub-assemblies can be combined with each other according to the requirements of specific vehicle programmes. The body components developed in the InCar project show an intensive use of advanced high-strength steels, hot forming technology, tailor welded blanks and steel sandwich materials. Some examples of the most innovative InCar solutions are as follows:

- Tubular design concept of the front longitudinal members showing the potential for cost savings and for a weight reduction of up to 9%.
- Hot formed body side reinforcement making use of a tailored tempering process to increase the failure strain in areas subjected to severe deformations in a side impact.
- Six different B-pillar concepts offering up to 22% weight reduction by the application of a new hot forming steel in combination with tailored tempering in the lightest version of the assembly.
- Application of a steel/plastic composite panel as a firewall for improved acoustics, suitable for resistance spot welding by adding electrically conductive particles to the viscoelastic core layer.
- Magnesium roof module offering 62% weight reduction.
- Advanced side door concept combining several parts in an inner panel including the side-impact beam as well as the hinge and lock reinforcements – up to 11% weight reduction at costs equal to reference.
- Bonnet with an outer panel made of a stiffness-optimised steel sandwich blank and meeting the pedestrian impact requirements as a passive system, accompanied by 11% reduction in weight (Hoffmann, 2009, ATZ, 2009).

The InCar project thus shows that in spite of the already widespread implementation of high-strength and ultra high-strength grades in series production, the classic body material steel still offers further potential for weight reduction. At the same time, it presents an interesting approach to meet the demand for flexible solutions in material usage responding to different priorities in terms of weight reduction, cost-efficiency and functionality.

Although SuperLIGHT-Car and InCar represent only a small extract from all the relevant research and innovation activities with regard to material usage in automotive bodies, these two projects already reveal that there is no simple answer to the question what the future material usage in this field might be. Most probably, the 'competition of materials' will go on and result in more and more diversified solutions subject to individual manufacturers' requirements and ever changing boundary conditions.

9.5   References

Adam H (2008), 'Cost-effective weight reduction with steel', Aachen Body Engineering Days 2008, Aachen, 23–24 September.
AISI (2006), 'Steel bumper systems for passenger cars and light trucks', Revision No. 3, American Iron and Steel Institute.
ATZ (2009), various articles, Automobiltechnische Zeitschrift, special edition 'The ThyssenKrupp InCar project'.
Bassi C (2009), 'Novelis multi material fusion technology for innovative car body solutions', Aachen Body Engineering Days 2009, Aachen, 22–23 September.
EAA (2010), 'Stiffness relevance and strength relevance in crash of car body components', European Aluminium Association.
Eichhorn O (2009), 'The new Polo – high quality and dynamic', Aachen Body Engineering Days 2009, Aachen, 22–23 September.
Eckstein L (2009), 'Kunststoffverarbeitung im Fahrzeugbau', Aachen, Institut für Kraftfahrzeuge, RWTH Aachen University.
Eckstein L (2010), 'Structural design of vehicles', Aachen, Institut für Kraftfahrzeuge, RWTH Aachen University.
EUR-Lex (2003), 'Directive 2003/102/EC of the European Parliament and of the Council', www.eur-lex.europa.eu, 2003.
EUR-Lex (2007), 'Regulation (EC) No 715/2007 of the European Parliament and of the Council', www.eur-lex.europa.eu, 2007.
EUR-Lex (2009), 'Regulation (EC) No 78/2009 of the European Parliament and of the Council', www.eur-lex.europa.eu, 2009.
Fidorra A (2010). 'The Art of Progress – Audi – The new A8', 12th International Conference EuroCarBody 2010, Bad Nauheim, 19–21 October.
Floeck S and Pfestorf M (2008), 'The body in white of the new BMW 7 series', Aachen Body Engineering Days 2008, Aachen, 23–24 September.
Goebbels R (2010), 'Analyse sekundärer Gewichtseinsparpotenziale in Kraftfahrzeugen', Automobiltechnische Zeitschrift, 12/2010.
Goede M (2009a), 'SuperLIGHT-Car project – an integrated research approach for lightweight car body innovations', Innovative Developments for Lightweight Vehicle Structures, Wolfsburg, 26–27 May.

Goede M (2009b), 'The SLC car body – innovative lightweight developments of SuperLIGHT-Car', Aachen Body Engineering Days 2009, Aachen, 22–23 September.

Gruenn R (2004), 'Die Kunst des Karosseriebaus', *Automobiltechnische Zeitschrift*, special edition 'Der neue BMW 6er', 36–52.

Hartmann B, Schmitt F and Eckstein L (2010), 'Lightweight measures in electric vehicles', *Automobiltechnische Zeitschrift*, 11/2010.

Hirsch J (2010), 'Recent aluminium alloy development for automotive applications', Strategies in Car Body Engineering 2010, Bad Nauheim, 15–16 March.

Hoffmann O (2009), 'InCar – innovative solution kit for body structures optimized in terms of cost, weight and function', Aachen Body Engineering Days 2009, Aachen, 22–23 September.

Hoffmann O (2011), 'Steel – the modern car body material with a perspective', Materials in Car Body Engineering 2011, Bad Nauheim, 11–12 May.

INF (2010), homepage of InflationData.com. Available from: http://www.inflationdata.com

Koehr R (2009), 'The lightweight body of the Porsche Panamera', Aachen Body Engineering Days 2009, Aachen, 22–23 September.

Kranz U (2010), 'Wir haben in Kühlschränke und Badezimmer geschaut', *Automotive Agenda*, 06, 72–76.

LME (2010), homepage of London Metal Exchange. Available from: http://www.lme.com

Macha U (2008), 'Development of an aluminium door structure', 6th European Automotive Conference on Hang-On Parts, Bad Nauheim, 17–18 November.

Pech C (2009), 'Realisation of an uncompromising sportscar concept – the body of the Mercedes SLS AMG', Aachen Body Engineering Days 2009, Aachen, 22–23 September.

Schenn A (2007), 'Der neue BMW X5 – Agilität, Variabilität, Exklusivität', Tag der Karosserie, Aachen, 8 October 2007.

Stauber R C (2009), 'Automotive materials – requirements and trends', Aachen Body Engineering Days 2009, Aachen, 22–23 September.

Weber M (2005), 'Ganzheitlicher Leichtbau mit innovativen Stahlprodukten, Verfahrenstechnologien und Karosseriekonzepten', VDI-Seminar 'Leichtbau mit metallischen Werkstoffen', Bremen, 26–27 April.

Weber M (2007), 'Ganzheitlicher Leichtbau mit innovativen Stahlprodukten, Verfahrenstechnologien und Karosseriekonzepten', VDI-Seminar 'Leichtbau mit metallischen Werkstoffen', Bremen, 12–13 June.

Wohlecker R (2007), 'Determination of weight elasticity of fuel economy for ICE, hybrid and fuel cell vehicles', SAE Paper 2007-01-0343, Detroit, Society of Automotive Engineers.

10

Advanced materials and technologies for reducing noise, vibration and harshness (NVH) in automobiles

T. BEIN, J. BÖS, D. MAYER and T. MELZ,
Fraunhofer Institute for Structural Durability and System
Reliability LBF, Germany

Abstract: The automotive industry is facing the problem more and more of reducing the weight of vehicles but guaranteeing an equivalent level of comfort in terms of noise, vibration and harshness (NVH). To overcome these contradicting requirements traditional design and material choices must be revisited. Besides advanced passive material active systems or smart concepts are being increasingly considered for the NVH optimization of vehicles. Therefore, this chapter addresses different passive and active measures for NVH control with a focus on smart structures. After a general discussion of the NVH problems in automotive engineering, some principal passive and active measures are described. On this basis passive and active measures are presented in different depth from general aspects over specific concepts to some selected applications.

Key words: noise and vibrations, active and passive measures, smart structures.

10.1 Introduction

Progress in material research is one of the driving forces behind innovative ideas not only for vehicles but in all fields of technology. This means that the availability of new materials, their manufacturing technologies, and their implementation in products is one of the crucial factors for economic success. New materials and their manufacturing must be equally in tune with sustainability, economic viability, safety, and a high level of component and system reliability. Beyond this, material development will probably concentrate more on materials and components with multifunctional characteristics and the way they are integrated into systems. In the next decades, the focus on material research for automotive applications will be driven by the demand for lightweight engineering, CO_2 reduction and energy efficiency as well as for component and system reliability. Furthermore, information technology will be developed to greater integration densities, and peripheral sensor and actuator applications are being introduced with highly integrated sensor and actuator functions. Since raw materials are limited and environmental contamination

254

(e.g., emissions) must be decreased, a stronger emphasis must be given to material recycling and the substitution of hazardous substances. Moreover, economic aspects will remain a decisive factor in materials selection for mass products in cars because the material costs and their impact on production processes form a substantial share of the overall product costs. Therefore, research will focus on material solutions that are cost-efficient or show the potential of becoming economically viable in the future.

Today's cars represent a complex compromise between contradictory requirements with regard to safety, exhaust emissions, noise, performance, comfort, and price. However, since it is widely recognized that the quality of life, particularly in the urban environment, is heavily influenced by air and noise pollution resulting from road traffic, one of the top priorities for car manufacturers is the reduction of noise and emissions from vehicles. As noise is considered to be one of the worst environmental pollutions worldwide, noise abatement concepts are demanded for vehicles of all kinds to protect people. In this regard, the principal vehicle manufacturers in Europe have unanimously agreed to adopt an integrated approach that has the development of more fuel-efficient power trains and weight reduction of the vehicle body as cornerstones.

In the past, the different attributes in vehicle design such as weight, efficiency, vehicle dynamics, safety as well as noise, vibration and harshness (NVH) have not been considered in a coherent way necessitating the introduction of additional, mostly heavy and costly measures to meet all requirements and customers' expectations. As an example, the structural weight of cars has been significantly reduced over the last decades but the total mass has increased due to higher safety, comfort, and functionality requirements. Nowadays, all attributes – of which NVH is the most relevant one – are considered in an integrated approach to optimize the vehicle on a system level. However, overcoming contradicting requirements arising from lightweight design, safety, and NVH remains a challenging task. It will be important to introduce new, passive or active material systems that enable a multi-material, lightweight design enhancing emission, safety, comfort, life time, and reliability standards at the same time. In addition, the automotive industry is also forced to implement even higher functionality into their vehicles in order to meet customer needs and to stay competitive. Both sustainable development and the addition of higher functionality in products are to be complemented by an NVH-optimized product design. It is well known that lightweight structures tend to vibrate more easily and radiate noise on a higher level demanding special effort to compensate for these higher noise levels. This often requires additional mass, which, in turn, nullifies the lightweight design. In the future, noise-optimized vehicles can only be achieved by combining passive and active noise and vibration abatement strategies based on smart materials that allow a lightweight design and higher functionality with respect to noise and vibration control. It must be stressed that the same technology can also be applied to a concurrent lightweight design by controlling the structural properties that impact on fuel consumption, exhaust emission, or safety aspects of vehicles.

For example, today engine downsizing or electrification of the drive train represents the most direct and cost effective approach to improving fuel efficiency in road vehicles, pushing the technological development of drive trains to new levels in order to ensure significant reductions of the impact on the environment while still providing acceptable levels of performance and vehicle 'fun-to-drive'. However, this new generation of drive trains, while being ideally suited to city vehicle applications, can result in a perceivable degree of deterioration in terms of noise and vibration when compared with the vehicles currently on the market. Correspondingly, in order to ensure that such vehicles achieve customer acceptability and the level of market penetration required to really make the difference in terms of reduced emissions from road transportation, particularly in the extremely important urban environment, it is imperative that robust solutions are developed that are viable, and sustainable. However, improvement of vehicle noise and vibration without affecting other performances is proving to be extremely difficult if not impossible with state-of-the-art technology. Frequently, new technologies in the fields of smart materials and active control provide potential solutions but have only been proved in the laboratory.

The aim of this section is to provide a brief overview of concepts addressing the NVH optimization of vehicles and to give an outlook on new trends focussing on material-related issues, in particular on smart solutions. Fundamentals on noise and vibrations are not being addressed.

10.1.1 Problems and challenges in current automotive design related to NVH

When speaking of NVH in the context of automotive vehicle design, there is no clear definition available as to which phenomena and frequency ranges are being addressed by noise, vibration, and harshness. Depending on how the phenomena are being perceived by the occupant or environment they are considered either as noise, vibration, or harshness. However, most of the problems are being received either as vibration or noise whereas the frequency range between 100 Hz and 500 Hz is nowadays the most challenging one. In this frequency range, airborne and structure-borne sound phenomena are of equal importance. Therefore, in the following discussion of NVH in automotive design only noise and vibration aspects will be considered with a slight emphasis on noise.

Travel pleasure nowadays is more and more frequently synonymous with individual wellness and performance feeling during transfer phase, which mainly depends on the control of ambient parameters and on the perception of noise and vibration levels. With regard to automotive engineering, the vibro-acoustic design deals with noise and vibration experienced by passengers inside the cabin and perceived exterior noise. It is largely concerned with the time signal history and frequency spectrum characteristics of major airborne and structure-borne sound/noise sources.

It is also important to understand that annoyance is not only a matter of the energy content but is being somewhat affected by subjective impression. This means that frequencies with same overall energy levels are sometimes not equally disturbing and experienced by persons differently. For this reason, it can be more effective to control particular harmonics or act on a specific frequency range only, for example by means of optimized tuning of active or passive control devices rather than reducing sound pressure or acceleration peaks over a broader frequency range aiming to simply make the environment quieter. These psycho-acoustic aspects of NVH increasingly influence the design of the interior noise and vibration levels, the choices of materials and the noise and vibration abatement strategies. For this, it is fundamental to understand how materials and noise and vibration abatement measures can be developed and adapted to the demanding and specific needs in automotive engineering taking while at the same time into account psychophysic effects, discomfort sensation, speech intelligibility, and privacy inside the vehicle interior.

Within a vehicle many components contribute to the vibration levels experienced and to the overall emitted noise between 60 dB(A) and 70 dB(A) of a vehicle (Fig. 10.1). Having different dominant noise and vibration sources of the same order (e.g., within 4 dB(A)), the treatment of only one source will not affect the overall vibration levels and radiated noise. Contrariwise, a previously masked noise or vibration source could become dominant and become more annoying than the treated noise or vibration source. However, in many cases the noise and vibration sources cannot be treated at all so that the noise and vibration propagation within the vehicle must be dealt with. As many different noise sources exist within the vehicle, even more different transfer paths for noise and vibrations need to be treated (Fig. 10.2). Again, it will not be sufficient to eliminate just one transfer path but all relevant ones. In order to achieve an overall noise and vibration reduction for vehicles, all noise and vibration sources and their transfer paths to radiating components must be treated simultaneously and in a holistic approach.

The majority of new vehicles in the next 15 to 20 years will still be equipped with advanced internal combustion engines but having reduced overall weight to meet the emission regulations. Due to the consequent lightweight design, the problem of multiple noise and vibration sources and transfer paths will become increasingly challenging with the advanced multi-material design of the vehicle body and envisioned flexibility and

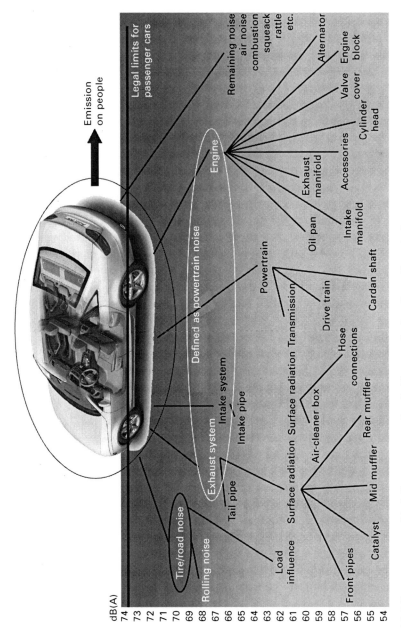

10.1 Components with influence on noise (with permission from Porsche).

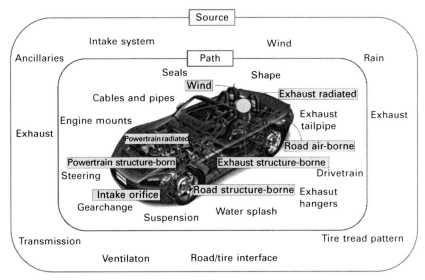

Intake system

Source

Wind

Ancillaries

Path

Rain

Seals

Shape

Wind

Cables and pipes

Exhaust radiated

Engine mounts

Exhaust
tailpipe

Exhaust

Exhaust

Powertrain radiated

Road air-borne

Powertrain structure-born

Exhaust structure-borne

Steering

Drivetrain

Intake orifice

Road structure-borne

Exhasut
hangers

Gearchange

Water splash

Suspension

Transmission

Tire tread pattern

Ventilaton

Road/tire interface

10.2 Transfer paths within a vehicle (with permission from Brüel and Kjaer).

modularity of the vehicle drive train. The dominant sources and their quality (e.g., frequency content) will vary from car to car, and additional transfer paths will occur. Although current vehicles already have a structure optimized with respect to weight and required performance, the weight can be further reduced without neglecting the resistance, crash, and fatigue performance of the structure. However, this will result in a poor performance in terms of noise and vibrations and increases both the structure-borne and airborne sound transmission. The most common solution used in the past was adding mass to the system for damping and acoustic treatments, which is not useful when the goal is weight reduction. Consequently, new approaches based on advanced passive and/or active materials must be applied.

With the upcoming electrification of the drive train the challenges for noise and vibration abatement will even rise. Although one of the dominant noise and vibration sources, the internal combustion engine, will be eliminated, its masking effect will also be eliminated. This way, the rolling noise or in general noise and vibrations excited by the road-tire interaction and aerodynamically excited noise and vibration will become dominant with different frequency content and different transfer paths. Well-established solutions for noise and vibration abatement might then no longer be applicable. Furthermore, sound engineering as part of NVH engineering will become more important with increasing electrification of vehicles. The psycho-acoustic impression of vehicles is, and will be, a purchase decision and is strongly associated with the brand name. The difference of the vehicle sound will perish with an electric drive train. Consequently, new ways based on materials and active measures must be found to create a well perceived typical sound profile for each brand. Ideally, the electric car will sound like a car driven by internal combustion. In addition, the low exterior noise of electric vehicles at low speed is a serious safety issue since pedestrians and bikers (in general vulnerable road users) cannot hear the vehicle any more. Again, a sound profiling is needed to support the identification of an approaching vehicle. However, since sound engineering is a very specific aspect of NVH it will not be addressed within this chapter.

10.2   General noise, vibration and harshness (NVH) abatement measures

The introductory discussion has indicated how complex noise and vibration treatments can become. Since many parameters (e.g., geometry or material) often are pre-defined by other requirements than NVH or sound (e.g., structural durability, crashworthiness, function, or visual appearance) the optimization space is limited with partly contradicting requirements. Particularly, a consequent lightweight design is in most cases opposing the requirements for high NVH comfort. If the noise and vibration aspect is not taken into account right from the beginning of a new product design cycle the NVH or sound targets are often not met and require subsequent improvements with penalties in weight, function, and costs. Particularly, in the mid-frequency range with its coupled vibro-acoustic phenomena, design neglecting the NVH aspect and focussing only, for example, on structural durability or safety, could negatively impact the vibrational characteristics and the radiated noise. Changing the stiffness, damping, or mass of a structure to deal with vibrations may change the structure-borne sound function altering the characteristics, for example, in terms of frequency content or increasing the sound power of the radiated noise. Local treatments can also alter the vibro-acoustic transfer paths to areas where noise and vibrations are even more critical. Adding damping may change the dimensionless radiation efficiency – depending on how it is done and which material is used – or shifting the resonance frequency to lower values due to increased mass resulting in increased vibration amplitudes. Minimizing the vibrations and radiated noise or designing the fundamental frequency or the radiated sound of a structure is therefore a highly complex optimization problem where all parameters used in the structural design of components must be taken into account.

In current automotive engineering different technical approaches are developed to optimize the NVH characteristics of vehicles. In principle these approaches can be divided in primary and secondary measures. Primary measures are aimed at prevention or abatement of the excitation, development, dispersion, and radiation of structure-borne sound. This is realized as closely as possible to the actual noise and vibration source (driving point) and is very efficient since further measures away from the source can be omitted or reduced. Secondary measures are those methods that are applied subsequently to control or reduce already generated vibrations or radiated noise. Both categories again can be divided into passive and active measures. While passive measures do not require additional energy in operation and use, active measures require energy during operation. Table 10.1 gives a general overview of the categories also listing some sample applications.

10.2.1 Passive measures

In general, passive measures for NVH consider the excitation, the structure-borne sound transfer function or the structure-borne sound efficiency and noise the radiation efficiency. As indicated in Fig. 10.3, the structure-borne sound function describes how the mean velocity of the radiating surface is related to the excitation force. The surface velocity determines the radiation efficiency via the radiated sound power.

Measures on the excitation as well as on the structure-borne sound transfer function will impact vibration and noise at the same time. For noise it must be considered that in principle the noisiest source should be treated first since it is masking the other noise sources. Noise sources of lower importance do not contribute much to the overall sound power level. They should only be treated if these noise sources influence negatively the frequency spectra and characteristics of a sound.

Table 10.1 Different measures for NVH optimization

	Primary measures	Secondary measures
Passive measures	Ribbings, stiffeners, damping, isolation, vibration absorber, increasing the driving point impedance	Enclosures, acoustic insulation, sound attenuation, acoustic ceilings, sound barriers, vibration absorbers
Active measures	Active vibration control (AVC), active structural acoustic control (ASAC)	Active noise control (ANC)

Copyright Fraunhofer LBF.

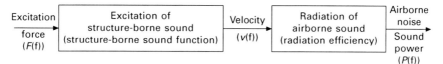

10.3 Block diagram for the fundamental equation of machine acoustics.

Excitation

Reducing the characteristic and level of the excitation is the most efficient way to control noise and vibrations. The vibrational response is proportional whereas the radiated sound power is proportional even to the square of the excitation forces. However, the modification of the excitation forces is often not possible since they are inherent to the underlying process or function of the component. Nevertheless, with respect to noise the underlying process and the function of the component should be designed in such a way that impulsive and shock-like excitations are avoided since those lead in general to highly disturbing noise. With respect to vibrations, any excitation forces at and near resonance frequencies of the excited structures must be avoided. Some rules for reducing excitation forces are listed below (Storm and Hanselka, 2008; Hanselka and Nordmann, 2011):

- continuous slope of the force progression
- balancing of rotating parts
- avoid tolerances between moving parts
- high surface quality of parts gliding on or rolling off each other
- low dimension and form tolerances (high manufacturing precision)
- avoiding impulsive and shock-like excitation
- avoiding excitations near or at resonances.

Most of the above rules are very cost-intensive and cannot be implemented in cost-sensitive products like cars.

Structure-borne sound function

Measures reducing the structure-borne sound function impact both noise and vibrations. However, reducing the structure-borne sound function often leads to a higher radiation efficiency (see below) for which reason both parameters must be considered simultaneously while aiming to lower the level of the acoustic transfer function. Nevertheless, the reduction of the structure-borne sound function is in many cases much higher than the increase of the radiation efficiency so that an independent treatment of the structure-borne sound function is a valid approach. Furthermore, the structure-borne sound function is more easily controlled by structural design than the radiation efficiency. Since flexural modes on the surface are the dominant source for radiated noise, large amplitudes of the structure-borne sound function by a given excitation should be avoided or reduced. The design should be made in such a way that the force flux is not directed towards surfaces easily radiating noise but rather kept in small, massive, and stiff designed parts of the structure. It is obvious that lowering the amplitude of the structure-borne sound function also leads to lesser vibrations.

Some rules for reducing the structure-borne sound function are listed below (Storm and Hanselka, 2008; Hanselka and Nordmann, 2011):

- forces should preferably act on compact, straight-walled structures
- separation of function: sound radiating surfaces designed very flexible as poor radiators and decoupled – in terms of structure-borne sound – from the load-carrying structures; driving point of excitation at the load-carrying structure
- increasing the impedance at the driving point of excitation (e.g., ribs and stiffeners near the driving point; irregular distribution of ribs; ribs high rather than wide)
- use of materials with a high density and high Young's modulus or increase of wall thickness (contrary to the lightweight aspect)
- rabbets and joints with high damping
- surface area preferably small.

Radiation efficiency (only for noise)

A reduction of the radiation efficiency by structural design is in most cases more complex and less efficient than reducing the structure-borne sound function. Nevertheless, in specific cases it might be useful as a complementary measure.

Some rules for reducing the radiation efficiency are listed below (Storm and Hanselka, 2008; Hanselka and Nordmann, 2011):

- the design of components should be rather compact, therefore radiating like a monopole; the radiation efficiency at lower frequencies decreases
- choose low wall thickness for structures with plate-like walls exploiting the acoustic short-circuit (could contradict requirements with respect to vibrations increasing the structure-borne sound function)
- for covers and liners not sealing a cavity air-proofed and not used as sound-barriers use perforated sheets with more than 30% area fraction of holes.

10.2.2 Active measures

The underlying technology for active vibration control (AVC) and active structural acoustic control (ASAC) are so-called smart structures or adaptive structural systems (adaptronics)(Bein *et al.*, 2005). Both terms are used synonymously. Based on intelligent material systems, adaptronics is introduced as an innovative, new cross-sectional technology for the optimization of structural systems. Adaptronics integrates additional functionality by combining conventional structures with intelligent material systems, which extend the classic function of load-bearing and shape-defining structures to sensing and actuating capabilities (see Fig. 10.4).

In connection with suitable adaptive controller systems, adaptive structural systems can optimally adapt to their respective operational environment. Intelligent material systems themselves are built up from passive, conventional lightweight materials and so-called transducer materials with actuating and sensing properties (Fig. 10.5). Not all transducer materials shown in Fig. 10.5 are of interest for noise and vibration control. Basically piezoceramics,

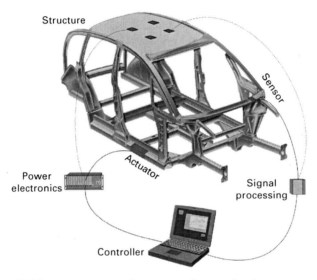

10.4 Smart structures for automotive applications.

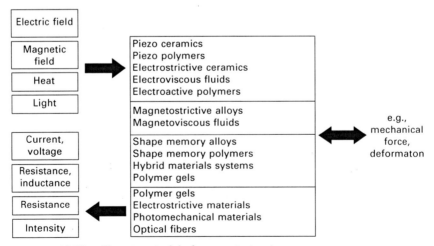

10.5 Intelligent materials for smart structures.

magneto- or electrorheological fluids (MRF) and electro-active elastomers/polymers (EAE/EAP) are widely applied.

While mechatronics extends the functionality of the existing structure only in relation to a design point mainly by adding components, the core objectives of adaptive structure technology consist of the following:

- the active control of the structural dynamic characteristics of the overall system and
- the optimization of the structure by replacing structural components with intelligent material systems (effective in the sensor/actuator sense) to save mass and design space.

Adaptive or intelligent solutions are thus characterized by adding functionality with the highest degree of integration leading to a structure-conform integration of the sensor and actuator components and to an active control of mechanical properties while assuming mechanical load-bearing characteristics in the overall structure. They fundamentally expand the structure characteristics and permit the abatement of any disturbances at their source, in the transmission paths, and/or in sensitive areas of the system. In realizing smart structures, the competences from various fields in today's engineering sciences need to be combined as shown in Fig. 10.6.

Adaptive structural systems are divided into three classes defined by the level of energy required (Fig. 10.7). The passive system does not require any

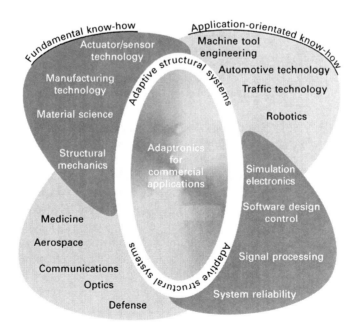

10.6 Multidisciplinary approach for adaptive structural systems.

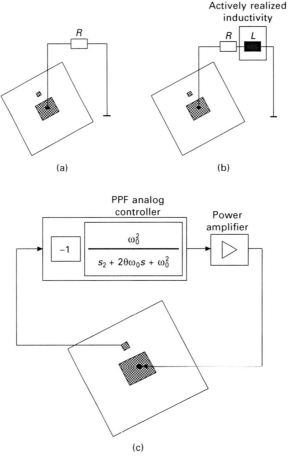

10.7 Adaptive structural systems (a) passive system; (b) semi-active system; (c) fully active system.

energy during operation. Nevertheless, it is still considered an adaptive system since it needs transducer materials and simulates mechanical functions by an electro-mechanical system. An example within this class is the so-called shunted piezoceramic. The second class is the semi-active system where only energy is needed to power the electronics of the system. Consequently, the power levels needed are low. Such a semi-active system is an electrical tuned vibration absorber based on piezoceramics. The most complex system is the fully active system where energy is needed to drive the actuators, sensors, and the control circuit.

While designing smart structures, the specific underlying problem must be considered for the selection of a suitable concept. Considering only NVH-related problems, basically AVC and ASAC concepts are pursued. Both are aimed at controlling the vibrational behavior of the structural components by controlling the mechanical properties (stiffness, damping), by minimizing the excitation leading to the vibration, or by applying destructive interference. The latter uses mostly planar actuator systems embedded in the construction material that are able to excite exactly the vibration modes to be controlled. The same planar actuators can also be used to add electro-mechanical damping. Figure 10.8 shows some examples where adaptronics is being considered within automotive engineering.

10.3   Selected concepts for noise, vibration and harshness (NVH) control

When considering NVH in the automotive design process, passive measures should be considered first. In general, they are well understood and are in terms of cost, production, and end-of-life issues preferable against active ones. However, the passive measures are already reaching their performance limits, and most likely future NVH regulation and customers' expectations towards comfort might not be satisfied by passive measures only. Therefore, selected concepts for both passive and active measures are described within this section.

10.3.1 Concepts for passive measures

As discussed in the previous section, passive measures include a variety of concepts ranging from controlling the excitation force to the design and to materials. With respect to materials, new advanced materials allow a new design with much better NVH characteristics than one can expect from the material alone if properly applied. However, the choice of material is in many cases made by requirements coming from the weight (lightweight), from the

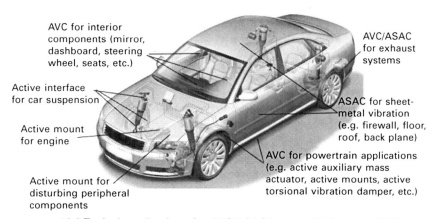

AVC for interior components (mirror, dashboard, steering wheel, seats, etc.)

AVC/ASAC for exhaust systems

Active interface for car suspension

Active mount for engine

ASAC for sheet-metal vibration (e.g. firewall, floor, roof, back plane)

AVC for powertrain applications (e.g. active auxiliary mass actuator, active mounts, active torsional vibration damper, etc.)

Active mount for disturbing peripheral components

10.8 Typical applications for AVC/ASAC in cars (Melz *et al.*, 2007).

structural durability, crashworthiness, manufacturing and production issues rather than constraints of its NVH characteristics. The latter is determined – among others – but by the bending stiffness, mass distribution, and damping of the structure as well as geometrical parameters such as wall thickness and surface area. These parameters are not independent from each other and are often influenced by the material's Young's modulus, material density and material damping but also by the way the structure is manufactured (e.g., welding, forming, etc.). Although advanced materials are one key issue within passive measures for NVH control, a complete overview would go beyond the limits of this section. Instead, only two aspects for passive measures related with materials in general are being discussed.

Materials

With respect to acoustics and particularly for the radiation efficiency of a structure, a low ratio of E/ρ – meaning a low stiffness at high mass – would be desirable but this is contrary to lightweight design and vibrations where a high ratio of E/ρ is needed. Consequently, the research on advanced lightweight materials such as porous foams, sandwich or composite materials considers more and more its acoustic properties such as transmission loss or absorption.

Composite materials such as fiber-reinforced plastics, plastic, cardboard, or aluminum honeycombs with metal, plastic, or non-woven skins provide an interesting compromise between weight and stiffness. They are often used for the production of trim components such as trunk load floors and parcel shelves or body parts such as spare-wheel pans. Fiber-reinforced materials are also used for the design of engine parts such as top covers and oil pans. The main disadvantage is that, being light and very stiff, they are typically efficient noise radiators when excited by structure-borne sound sources, and their coincidence frequencies tend to be very low. Such materials, therefore, provide an insulation performance that is inferior to the one of a homogeneous panel with the same mass. Hence, porous and composite materials were investigated within in European Integrated Project InMAR (InMAR, 2008). From the acoustic absorption point of view, porous materials, depending on their porous properties (geometry, size, distribution, etc.) possess sufficient performance in the medium to high frequency range, but they do not perform as well or are sometimes even useless in the lower frequency range. This problem can be overcome by adding smart materials to the porous materials, thus forming a hybrid material system. The active layer is acting as a loudspeaker controlling the volume in the porous material. This way, the absorption coefficient can be increased significantly in the frequency range between 200 Hz and 350 Hz as shown in Table 10.2.

Furthermore, advanced lightweight sandwich panels, which are poor sound

Table 10.2 Absorption coefficient on a hybrid porous material

Frequency (Hz)	Absorption coefficient	
	Porous material	Hybrid concept
200	0.07	0.23
225	0.08	0.24
250	0.07	0.25
275	0.09	0.25
300	0.09	0.30
325	0.08	0.22
350	0.09	0.25

Source: InMAR, 2008, INASMET.

insulators, were considered in InMAR (2008). Sandwich panels made of periodic soft core stripes were developed to improve the sound insulation. Up to 10 dB higher transmission losses (TL) were achieved using unidirectional or crosswise core stripe designs (Fig. 10.9).

Damping treatments

One of the key parameters for NVH control is the damping within the system where one can distinguish between the intrinsic damping of the material (material damping) and structural damping. The effect of structural damping is in most cases much higher than material damping. In automotive applications mostly extensional or constrained layer damping is applied. With extensional damping an additional layer of materials with a high material damping is bonded to the structure. While the structure is deformed or vibrates, energy is dissipated in the damping layer by compression or extension of the material. The thicker the layer the more damping will be introduced but with the penalty of additional weight. However, the added layer does not only increase the damping but it also affects the natural frequencies of the structure due to its additional mass. The damping layer consists in general of a viscoelastic material such as bitumen, rubber, epoxy, or PVA/PVC. Recently, damping layers are increasingly applied in a sprayable form (see Fig. 10.10(a)) to reduce the labor-intensive bonding of pre-fabricated damping pads (Fig. 10.10(b)). With respect to environmental issues there is a tendency to use water-based viscoelastic materials having a low density after curing (Christie, 2010).

Another approach is constrained layer damping where a viscoelastic layer is sandwiched between a stiff base layer (e.g., the body-in-white) and a constraining third layer. In this case, energy is dissipated by shear deformation of the viscoelastic layer. Current constrained damping treatments used in the automotive industry consist of either local applications of aluminum foils with a coating made of a soft and sticky viscoelastic material or of complete

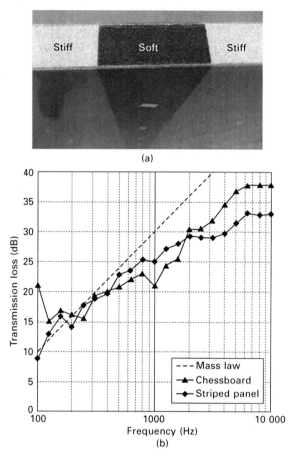

10.9 Cross-section of sandwich panel with a periodic soft core (a); transmission loss of this panel (b) (InMAR, 2008; work performed by University of Stockholm (KTH) and VTT).

panels made with laminated steel (two thin steel layers with an intermediate layer of viscoelastic material). The local solutions cannot be applied to beaded surfaces because the aluminum foil cannot be conformed to the geometry. Laminated steel can be beaded but it leads to a heavier structure since the total thickness of the steel must be higher than conventional to reach the same stiffness. Laminated steel is commercially available, e.g., as Bondal® (TKS, 2001), CoustiPlate (H&H, 2006), or QuietSteel® (MSC, 2010). These materials exhibit rather high dissipation factors up to $\eta \approx 0.3$ resulting in almost complete suppression of resonant vibrations in simple plates made of these materials.

However, it is an inherent property of viscoelastic materials that they have frequency and temperature dependent mechanical properties. This

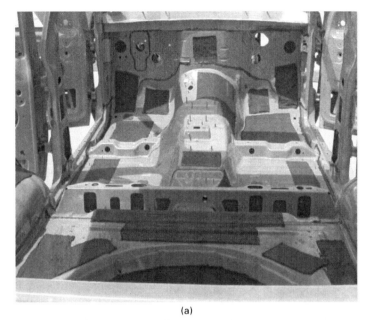

(a)

(b)

10.10 Sprayable damping treatment (Christie, 2010).

can affect the damping and limits the effective temperature and frequency range of the treatment (Gardonio, 2006). To overcome this problem hybrid concepts have recently considered combining passive layered damping with active elements (e.g., piezoelectric patches) to actively control the shear in the viscoelastic layer. With such a concept the loss factor validated with a cantilevered beam can be almost doubled (Gardonio, 2006; Kumar, 2009).

10.3.2 Concepts for active measures

If passive measures fail to achieve the desired performance with respect to NVH, the implementation of active concepts should be considered. In particular, active sound control concepts using loudspeakers as secondary sources are well developed in applications where the disturbance is tonal. Although the principle of controlling engine noise in cars was demonstrated some time ago, its commercial application has been slower than, for example, in the aerospace industry, because of the greater cost sensitivity of the automotive industry. Current commercial systems, for example, those used by Honda, however, include a feedback controller to control a resonance in the Accord station wagon (Sano *et al.*, 2001). A number of other companies are also currently investigating the production of systems for the active control of engine noise in the cabin, reducing costs by sharing loudspeakers, amplifiers, and signal processing with the audio system. The control of road noise introduces another level of complexity, due to the impossibility of using a well-defined control signal and to the broadband nature of rolling noise. Currently, there is no active noise control system in production for broadband rolling noise.

In general, active measures are more complex than passive ones but the performance gain is much higher. Some of the known concepts such as shunt damping or adaptive neutralizer can be combined with passive measures to hybrid systems utilizing the advantages of both approaches. In the following paragraphs some well-established active measures based on ASAC are described; however, many more concepts can be found in the literature.

Active interface

The principle of an active interface is shown in Fig. 10.11 (Bein *et al.*, 2008). Disturbances of any kind do not only act at the point of origin but are transmitted to anywhere within a structure. If the main transfer paths are known, an active interface can be placed in between, thus interrupting the energy transmission and isolating, from the dynamic point of view, the structure before and after the interface. Different concepts are used to control the structure-borne sound path depending again on the considered problem. Primarily active interfaces are used as active mounts or where bolted joints are applied. In both cases, either the impedance of the mounting or joint can be adjusted actively or a statically hard but dynamically soft spring can be designed.

The idea of active vibration cancellation is to detect the disturbance by

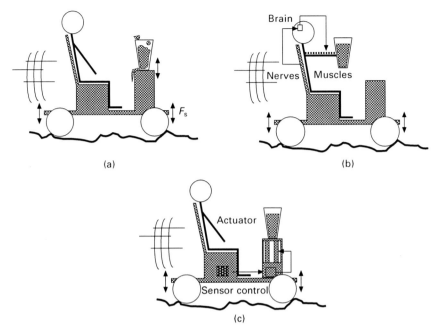

10.11 Principle of an active interface.

a sensor (e.g., an accelerometer), to process the sensor signal by a control algorithm, and to intensify the control signal by an amplifier in order to drive an actuator (e.g., a piezoceramic actuator) that induces a force or displacement in the target structure to counteract the disturbance.

In Fig. 10.12 the principle of active vibration cancellation is sketched (Bein *et al.*, 2008). As an example, a three degree-of-freedom (DOF) system is chosen that is excited through the supporting point. Moreover, the active interface is included as a force generator driven by a simple velocity feedback controller. According to Elliott (2008) the optimum force generated by the active interface is given by

$$f_{ai} = - S_M \cdot x_s \qquad [10.1]$$

where S_M is the mount stiffness and x_s the displacement of the base point excitation (according to Fig. 10.12). With this approach it is possible to significantly reduce the vibration at the natural frequencies in the transfer function, i.e., additional artificial damping is introduced into the system dynamics. In this example, the active interface is modelled as a simple ideal force generator. However, in reality it represents a complex assembly consisting of sensors, actuators, electronics, and mechanical devices to bear the loads occurring during operation or in misuse situations. Moreover, the interaction of the interface with the coupled structures must be considered during the design process. Hence, it is necessary to develop a design method

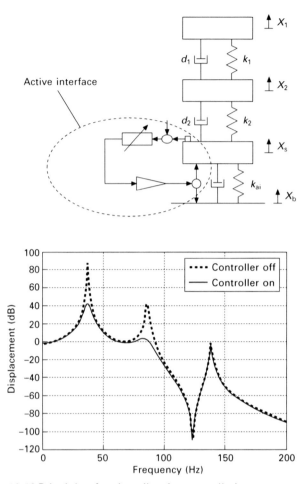

10.12 Principle of active vibration cancellation.

that emphasizes the system simulation as a means to estimate the performance of an active system as a solution for a given NVH problem.

An example for such an active interface is given in Fig. 10.13. The interface is designed for application in a car chassis, particularly mounted between the spring/damper unit and the car body. Thus, the interface is completely integrated into the load flux and must withstand all the loads that can occur during the lifecycle of a car. The aim of this placement is to actively decouple the car chassis from the car body in order to prevent noise induced by the road/tire contact to be transferred into the passenger's compartment.

As mentioned above the specific placement directly into the load flux and in a harsh environment leads to a complex design process that must take into account all possible stresses. Moreover, the active interface should not only be applicable particularly in the car chassis. It also serves as a basis

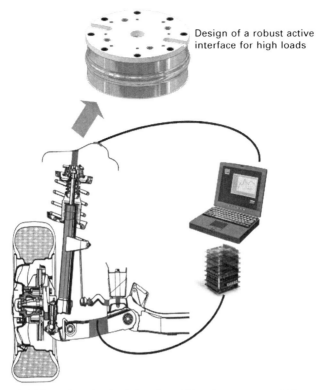

Design of a robust active
interface for high loads

10.13 Functional principle of a stiff active mount within a car
suspension system.

for the development of a vibration decoupling system for other applications
such as machine tools or engine mounts. Hence, it is designed to act in three
DOF.

Figure 10.14 shows the active interface as a CAD model and as a prototype.
It consists of eight piezoceramic stack actuators placed between two plates
and arranged in two stages. The two stages are connected in series; therefore,
it is possible to realize a very compact design. To handle lateral forces a
guiding membrane is used that functions as a flexural joint. It is stiff in-
plane and soft out-of- plane; thus, it does not influence the actuation of the
interface.

Shunt damping

A reduction of mechanical vibrations is also possible by means of shunt
damping technology. With the help of transducers a conversion of mechanical
to electrical power can be achieved (Fig. 10.15). This energy can either be
dissipated by a resistance or buffered in an electronic resonant circuit. The

 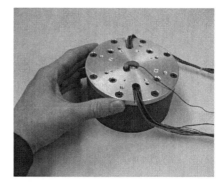

10.14 CAD model and prototype of the three DOF interface.

first increases the damping of only lightly damped structures whereas the latter behaves like a mechanical tuned vibration absorber.

In Fig. 10.16 an electromechanical equivalent is used to illustrate the circuit of a piezo transducer (Mayer *et al.*, 2009; Gardonio and Brennan, 2002). In this model only the mechanical stiffness k_P, the damping b_P, the electromechanical coupling coefficient α, and the electrical capacitance C_P are considered, whereas the actuator mass is neglected. The model parameters of a piezo stack can be calculated from the physical and geometrical parameters

$$k_P = E_P \frac{A_P}{l_P}, \quad C_P = \varepsilon_{33} \cdot \frac{A_P}{l_P} \cdot n_P^2, \quad \alpha = d_{33} \cdot n \cdot k_P \qquad [10.2]$$

In this equation A_P is the cross-section area, l_P the length of the actuator, and n_P the number of stacking layers connected both mechanically and electrically. The necessary physical parameters are the Young's modulus E_P and the piezoelectric material constants d_{33} and ε_{33}.

The electrical termination of piezoelectric transducers with a resistor and an inductor creates a resonant circuit. A similar behavior compared to a classical mass damper (Viana and Steffen, 2006) can be observed if the resonance frequency of the electric circuit and of the mechanical system is identical. This system can be described with an electrical circuit or by a mechanical system (Fig. 10.17).

The electromechanical equivalent describes the inductor as a mass, the resistor as damping, and the capacitance of the piezoceramic transducer as a stiffness. Optimal values for the resistor and the inductor can be calculated with Eq. 10.3 (Hagood and von Flotow, 1991).

$$L = \frac{1}{\omega_{n,s}^2 \cdot C_P \cdot (1 + K_{ij}^2)} \quad R = \frac{\sqrt{2} \cdot K_{ij}}{\omega_{n,s}^2 \cdot C_P \cdot (1 + K_{ij}^2)} \qquad [10.3]$$

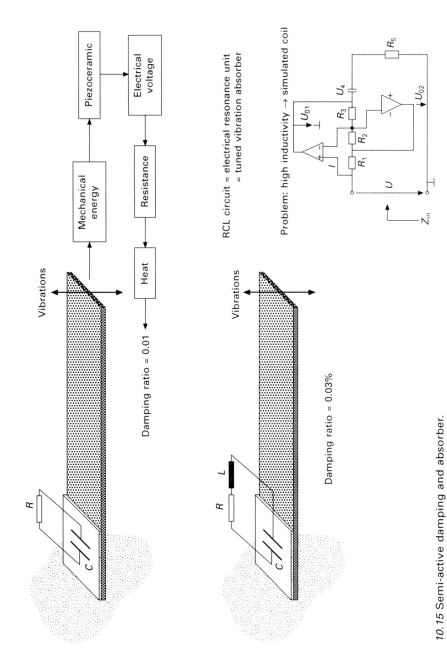

Piezoceramic → Electrical voltage

Mechanical energy

Resistance

Vibrations

Heat

Damping ratio = 0.01

RCL circuit = electrical resonance unit
= tuned vibration absorber

Problem: high inductivity → simulated coil

U_{01} R_3 U_4 R_5

R_2

R_1 U_{02}

U

Z_{in}

Vibrations

R L

C

Damping ratio = 0.03%

R

C

10.15 Semi-active damping and absorber.

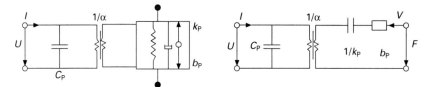

10.16 Electromechanical equivalent circuit of a piezo transducer.

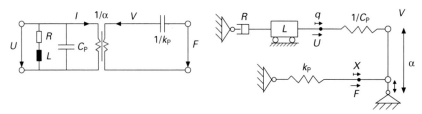

10.17 Electrical and mechanical model of a piezo transducer with RL circuit.

Generalized electromechanical coupling coefficient K_{ij} is defined as:

$$K_{ij} = \sqrt{\frac{\omega_{n,0}^2 - \omega_{n,s}^2}{\omega_{n,s}^2}}$$

[10.4]

Here, $\omega_{n,o}$ describes the natural frequency of the structure with open piezoelectric transducers and $\omega_{n,s}$ the natural frequency in the case of the transducers shorted. The advantages of vibration reduction with an RL shunt are that no electrical energy is used to drive an actuator, its small complexity and cost, and easy adaptability as well as good integration potential. Depending on the application and the value of the inductor either a passive component or a synthetic inductance, realized with an operational amplifier circuit, can be used. The latter gives the possibility to easily adapt the inductance and thus the tuning frequency during operation.

To illustrate the effect of a semi-active electromechanical vibration absorber based on a shunted piezoelectric patch actuator it is applied to a rectangular aluminum plate with a size of 340 mm × 300 mm × 2 mm. The vibration absorber is light and small, it can easily be tuned to a specific frequency by adjusting its resistance and inductance, and it requires only minimum external power supply (±15 V for the operational amplifier) and no control (Bös and Mayer, 2006). Based on optimal values for the resistance and inductance, an electrical circuit with a size of approximately 20 mm × 20 mm was assembled (see Fig. 10.18).

A generalized impedance converter was used as a synthetic inductance. Both the inductance and the resistance can be tuned so as to achieve a maximum vibration reduction. This maximum is reached for $R = 220\ \Omega$ and

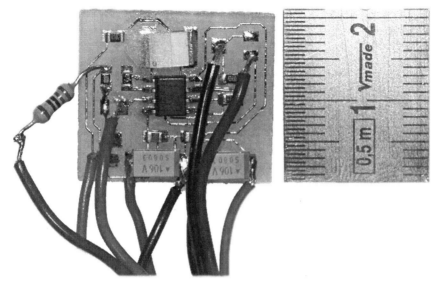

10.18 Electrical circuit of the semi-active vibration absorber.

$L = 9.6$ H. Although the measured capacitance of the piezoceramic patch matches the theoretical value very well, the real resistance and inductance differ greatly from their theoretically optimal values. Such discrepancies were also observed in (Hollkamp, 1994).

Figure 10.19 shows the test box made of acrylic glass (right). The above-mentioned rectangular aluminum plate forms the front panel of the box, which has a size of 340 mm × 300 mm × 320 mm. The acrylic glass has a thickness of 20 mm. A loudspeaker, which excites the aluminium plate, is suspended in the middle of the box. The piezo transducer, consisting of two piezo patches (see Fig. 10.19 left), is located in the middle of the aluminium plate on the interior of the test box.

The effectiveness of the semi-active vibration absorber is tested by means of scanning laser vibrometer and nearfield acoustical holography measurements. Both the loudspeaker excitation (see Fig. 10.19 right) and the impulse hammer excitation are used. Figure 10.20 shows the velocity spectrum at the plate's surface measured with the scanning laser vibrometer using the impulse hammer excitation. The velocity peak at 205 Hz drops by approximately 7 dB as the absorber is switched on. Interestingly, the velocity reduction is slightly larger for the loudspeaker excitation (−9 dB). Those measurement results even show the typical behavior of a vibration absorber, i.e., the amplitude is reduced at the tuning frequency of the absorber, but two new peaks are created in its vicinity. However, this effect is not visible for the impulse hammer excitation (see Fig. 10.20). The sound field radiated by the aluminum plate at its first vibration mode (loudspeaker excitation) was measured by means of nearfield acoustical

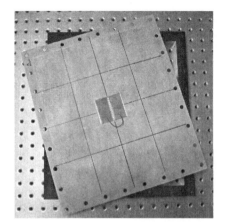 

10.19 Left: piezo patches attached to the aluminum plate; right: test box made of acrylic glass (with loudspeaker).

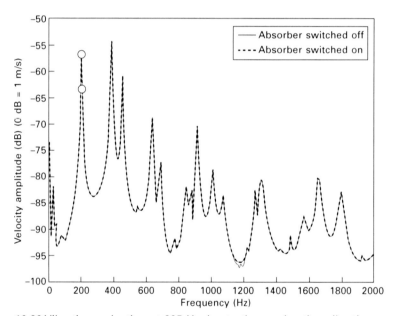

10.20 Vibration reduction at 205 Hz due to the semi-active vibration absorber.

holography. The sound pressure level is reduced by approximately 20 dB due to the semi-active vibration absorber.

Methods for reduction of forced vibrations

To reduce the effect of forced vibrations exciting a structure out of resonance, the most promising approach is to generate a counteracting force. In this section, three possible implementations of this idea are discussed: passive vibration neutralizers, adaptive vibration neutralizers (which both belong to the class of vibration absorbers) and an active system with the force being controlled by adaptive feed forward control (Mayer *et al.*, 2010).

The passive neutralizer consists of a simple single DOF oscillator (described by its main parameters mass m, resonance frequency ω_0 and damping coefficient θ) attached to a host structure, described by its mechanical input admittance $Y_M(s)$, as shown in Fig. 10.21(a). By taking the mechanical admittance of the oscillator at its base

$$Y_N(s) = \frac{1}{m_N} \cdot \frac{s^2 + s \cdot 2 \cdot \theta \cdot \omega_0 + \omega_0^2}{s^2 \cdot 2 \cdot \theta \cdot \omega_0 + s \cdot \omega_0^2}$$ [10.5]

the resulting mechanical admittance of the system can be derived by simply adding the impedances of the neutralizer and the host, and taking the reciprocal of the result, or by the regarding the whole system as some kind of control system (Fig. 10.22). The final result reads

$$Y(s) = Y_M(s) \cdot \frac{s^2 + s \cdot 2 \cdot \theta \cdot \omega_0 + \omega_0^2}{\begin{array}{c} s^2 + s \cdot 2 \cdot \theta \cdot \omega_0 + \omega_0^2 \\ + Y_M(s) \cdot (s \cdot \omega_0^2 + s^2 \cdot 2 \cdot \theta \cdot \omega_0) \cdot m_N \end{array}}$$ [10.6]

By adding two complex zeros to the mechanical admittance of the structure a vibration cancellation effect is caused at the resonance frequency of the neutralizer, thus the neutralizer should be tuned to the excitation frequency of the disturbance. As shown in Brennan and Kidner (1995), the performance

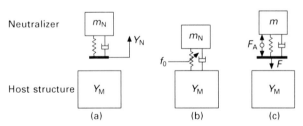

10.21 (a) neutralizer, (b) adaptive neutralizer, (c) inertial mass actuator attached to an arbitrary host structure.

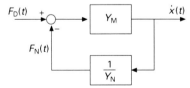

10.22 The neutralizer host structure system regarded as a control system.

of a neutralizer at the tuning frequency is directly proportional to the ratio of the neutralizer mass to the mass of the host structure, while the damping of the neutralizer reduces the achievable vibration reduction. For the case of the host structure being a pure mass, an example is shown in Fig. 10.23.

Obviously, the neutralizer effect is quite narrowband, i.e., the performance decreases drastically when the excitation frequency changes and does not match the resonance frequency anymore. As depicted in Fig. 10.23, increasing the neutralizer mass enhances the bandwidth of the vibration reduction. This is, however, contradictory to the goal of a lightweight structural design. An alternative is the application of a single DOF oscillator with an adjustable resonance frequency, which is tuned to the excitation frequency, so the zeros of the transfer function should track the disturbance (Fig. 10.21(b)).

An active system is designed by means of a dynamical actuator, which is capable of exciting a structure to vibrations. Here, an inertial mass actuator is chosen, since it represents just an attachment to the structure and is therefore comparable to the neutralizers discussed above. Similar to absorbers or neutralizers, the force is generated by an inertial mass; however, here the mass is actively excited by an internal actuator force F_A (Fig. 10.21(c)). The force generated by an inertial mass actuator can be calculated to

$$F(s) = - \frac{s^2}{s^2 \cdot 2 \cdot \theta \cdot \omega_0 \cdot s + \omega_0^2} \cdot F_A(s) \qquad [10.7]$$

which is proportional to F_A above the resonance frequency ω_0.

For an illustration of these three methods and a first evaluation, a very basic example is studied. The test system consists of a simple single DOF

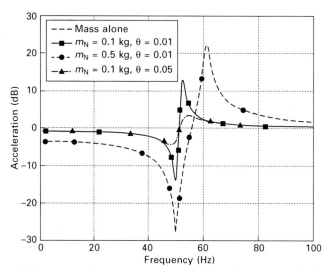

10.23 Acceleration of a 1 kg mass with a 50 Hz neutralizer attached.

oscillator with a resonance of 10 Hz excited by a harmonic disturbance force with a varying frequency. The frequency is chosen high enough in order not to excite the structure in resonance, which is a realistic assumption since most mechanical structures are designed in a way that known disturbance frequencies do not match the resonance frequencies of the system – which is actually the first and basic step towards vibration reduction with passive means. Thus, the disturbance signal is chosen as a swept sine starting at 40 Hz and increasing to 60 Hz. The rate of frequency increase is chosen slow enough in order to avoid transient effects from the adaptation of filter coefficients or the resonance of the adaptive neutralizer. The excitation amplitude is set to 1 N.

The resonance of the passive neutralizer is chosen at 50 Hz. A relatively high mass of 0.5 kg and a low damping ratio of 0.01 should lead to a broadband reduction of vibrations. The adaptive neutralizer is supposed to track disturbance frequencies around 10% of the nominal frequency of 50 Hz, thus its resonance can be shifted from 45 Hz to 55 Hz. The mass is chosen at 0.1 kg and the damping ratio at 0.01. The active system consists of an inertial mass actuator tuned to 10 Hz with a damping ratio of 0.03 and a mass of 0.1 kg.

Since the system parameters vary with time in the case of the adaptive neutralizer, a transient system simulation is set up. As a measure for the performance, the RMS value of the host structure acceleration is recorded, while the displacement of the neutralizer's mass (or that of the inertial actuator) is also monitored as a measure for the control effort.

As shown in Fig. 10.24 (a), the adaptive and the passive neutralizer show a similar performance, while the passive one has a mass five times higher than the adaptive one. The drawback of the adaptive neutralizer is, however, that in the chosen example the disturbance matches the resonance of the neutralizer-host system around 57 Hz after the tracking of the neutralizer stops at 55 Hz due to the chosen restrictions. Thus, an adaptive neutralizer should always be designed in a way that it can be adjusted to the maximum occurring disturbance frequency in order not to deteriorate the vibration reduction.

The active system shows a much better performance than the other systems, but at the expense of an external actuator force having to be provided. Obviously, both the inertial mass actuator and the adaptive neutralizer show similar displacement RMS values (Fig. 10.24(b)), while the adaptive neutralizer shows quite less vibration reduction. The reason is that the chosen filtered × least mean squares (FxLMS) method is a feed-forward method, which is able to cancel the error signal completely with a moderate amplification of the reference signal. In contrast, a mechanical neutralizer represents a feedback configuration, which needs a quite high gain in order to reach a good disturbance cancellation. However, the gain is restricted due to the physical quantities of neutralizer mass and damping.

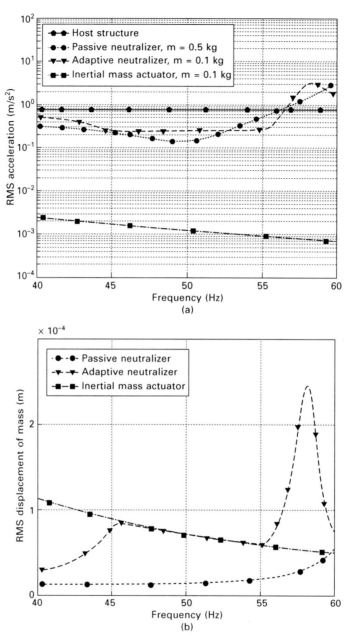

10.24 Acceleration of the host structure (a) and displacement of the mass of the vibration control system (b).

10.4   Applications

Since many applications for passive measures can be found in the literature only concepts by means of active measures are discussed in this section. Typical applications for active systems in automobiles focusing on AVC/ASAC are:

- soft active mounts as engine supports (Fursdon *et al.*, 2000)
- stiff active mounts (interface) to reduce road-tire-contact induced vibration (Matthias *et al.*, 2005)
- active struts to reduce torsional vibration within convertibles (Kalinke *et al.*, 2001, Kalinke and Gnauert, 2002),
- active add-on systems such as adaptive absorbers or auxiliary mass actuators (Melz, 2001) or
- in-plane actuators to reduce sheet-metal vibration of firewalls, roofs or windshields (Schmidt *et al.*, 2003).

These systems can be active or semi-active. Examples of commercially available semi-active systems are 'adaptable' car suspension damping devices which are realized, e.g. as controlled mechatronic CDC (continuous damping control) systems (ATZ, 2004), used within, for example, the Lancia Thesis, VW Phaeton, VW Touareg, Audi A8, Porsche Cayenne and Opel Vectra or semi-active engine mounts. Another recent example exploiting intelligent materials is the 'magnetic ride' designed for the current Audi TT (Krix, 2006). An even more challenging approach was developed by the Bose Corporation utilizing electromagnetic actuators replacing conventional car suspension systems (Goroncy, 2005). This active suspension system focusses on low frequency driving dynamics and has not yet been integrated within series-production cars. Stiff active interfaces are particulary interesting for higher dynamics up to the vibro-acoustic frequency range. An exemplary market need for an adaptronic solution results from the increasing application of run-flat tires – first introduced within luxury cars. These are stiffer than conventional tires, thus degrading the NVH characteristics within the passenger's compartment.

Semi-active soft engine mounts have also been investigated which make use of, for example, electromagnetic systems or even MRF (Marienfeld *et al.*, 2002, Janocha, 2006). Active mounts focus on integrating pneumatic, electrodynamic, and electromagnetic actuators (Matsuoka *et al.*, 2004). One current commercial system is integrated within the Jaguar XJ6 TDVi (Fursdon *et al.*, 2000) (Fig. 10.25).

In addition to the trend to increasingly use soft active engine mounts, there is an increasing importance to optimize the NVH characteristics of chassis. An example of a semi-active mechatronic system for chassis control based on electro-hydraulics was investigated in Gruber *et al.* (2003), whereas various

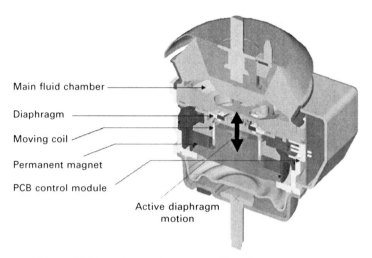

Main fluid chamber

Diaphragm

Moving coil

Permanent magnet

PCB control module

Active diaphragm
motion

10.25 Avon VMS active engine mount (Fursdon *et al.*, 2000).

active systems recently focused on AVC for convertibles, e.g., (Kalinke and Gnanert, 2002).

10.4.1 AVC of torsional vibration in convertibles

The dynamic stiffness of the car body predominates the driving comfort of a passenger car in the low frequency range (<30 Hz). The torsional mode is especially relevant for the driver's comfort impression. For convertibles the corresponding natural frequency is lower than, for example, sedans due to the fact that no stiffening roof structure exists ($f_t \sim 20$ Hz for convertibles compared to $f_t > 50$ Hz for sedans). To prevent high vibration amplifications of the car body, e.g., caused by the engine, it is preferable to realize relatively high natural frequencies of the car body. A low natural frequency results in perceivable vibrations of the rear-view mirror, the seats, and even the steering wheel. To reduce these vibrations different kinds of passive means are usually used such as stiffening the body by additional sheet-metals, enlarging the cross sections of beams, applying struts in the car underbody, or even using the rear panel as an additional load carrying part. Such means increase the total weight of a body-in-white of a convertible compared to a sedan by about 50 kg. Furthermore, heavy passive absorbers are added to reduce disturbing vibrations by typically 10–20% at the natural frequency. Typical system weights range from 8.5 kg to 14 kg in four seaters (Kalinke *et al.*, 2001; Kalinke and Gnauert, 2002). Of course, another important design rule is to prevent natural frequencies of subsystems coinciding, such as suspension, chassis, engine mounting, and car body. However, even with all these means the dynamic stiffness of a sedan cannot be reached.

One adaptronic approach to reduce torsional vibrations is to integrate actuators within the diagonal stiffening struts of the car underbody (Fig. 10.26). Different actuators have been investigated, e.g., piezoceramic stacks, hydraulic cylinders, and hydraulic muscles. Control approaches such as adaptive feed-forward and feedback were implemented, and significant vibration reductions were achieved (Fig. 10.26). To commercialize this affirmative active concept, system cost, size, complexity and power consumption must be further reduced. To achieve optimized solutions, one predominant task must be the matching of the mechanical impedances of the active struts with the car body.

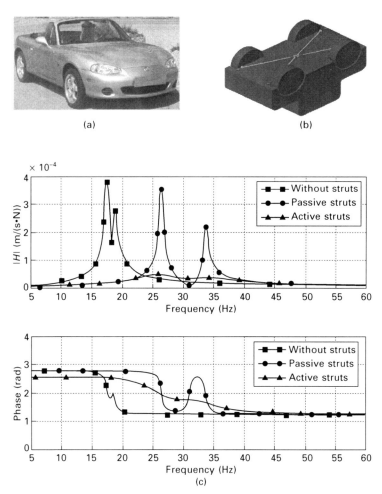

10.26 AVC-application in a convertible (a), concept of active struts (b) and achievable vibration reduction (c).

10.4.2 Active interface for an instrumental panel carrier

One of the first generations of an active interface simply used piezoceramic stack actuators as a shim in a bolted joint. In Fig. 10.27, such a simple setup is shown for the decoupling of the steering column from the instrument panel carrier (IPC). The IPC is designed in such a way that the fundamental frequency of the system IPC/steering column is higher than 40 Hz. This leads to a much heavier IPC than it is needed just for static and structural durability reasons. By decoupling the dominating mass of the steering column, the IPC can be designed much lighter (up to 50%) whereas the first resonances of the system are controlled with the actuators (Fig. 10.28).

10.4.3 Active interfaces for front and rear suspensions

Current car suspensions are designed as a compromise between car dynamics, comfort, cost, and weight. Typically, their operative frequency range is limited to the needs of low frequency car dynamics. Higher frequencies directly affect the passengers' NVH comfort. To optimize NVH characteristics, high damping rubber material and air springs for upper class cars are state-of-the-art whereas they reach limits with respect to increasing NVH and lightweight design demands of modern vehicles. Further NVH problems come from the new run-flat tires which are stiffer than conventional tires to prove fail-safe operation with no air pressure (Backfisch, 2006).

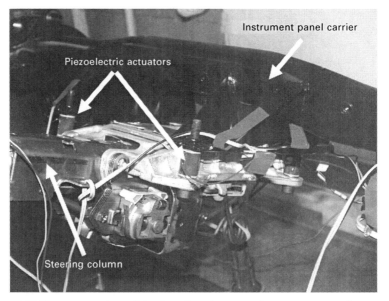

10.27 First generation of the interface between instrument panel carrier and steering column.

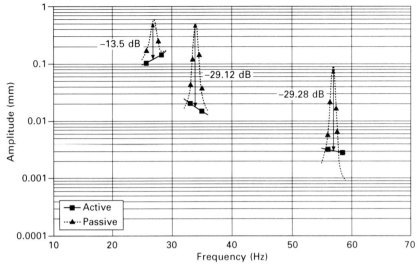

10.28 Comparison of the active and the passive instrument panel carrier.

A promising solution is to integrate stiff, high-load active mounts into the suspension system and, thus, actively prevent structure-borne noise from spreading into the car body (see also Fig. 10.13). Typical operative frequencies should range from approximately 30 Hz to several hundred Hz. As shown above, one design approach for such active interfaces is to integrate stiff piezoceramic stack actuators in an elastic housing and to design a robust, compact, and housed system. Using piezoceramic actuators ensures short response times and high active and passive load characteristics, which are necessary for a direct integration into the load transfer path.

A recent design study based on low cost actuators, which have been developed as a mass product for fuel injection systems (Blöcking and Sugg, 2006), is described in (Melz and Matthias, 2005). It provides a stroke of 70 μm and is designed for typical forces relevant within the front suspension. First tests have successfully been performed with loads of up to 18 kN in z-direction. Furthermore, it has three DOF (translation along z, rotation about x and y) although for this particular application only the translation is relevant. More recent versions are more compact and can be loaded even higher while providing an adapted reduced stroke. In the first experiments a significant broadband vibration reduction was shown (Herold *et al.*, 2005).

One interesting aspect of such stiff active interfaces for automotive applications is that they can be used for a variety of interconnection points within the car structure, e.g., rear suspension, power train, exhaust system, rear axles suspension, or motor. This is particularly interesting when properly combined with passive means such as elastomer or hydraulic supports.

10.4.4 Active engine mount

Among others, the reduction of noise transmission from the engine into the passenger compartment is a critical task for the improvement of the interior noise levels. Active engine mount systems based on the above mentioned active interface in series with the existing rubber mount can be used to implement an active vibration control system at one major source of the structure-borne noise transmission. The considered passenger car has mount systems at two positions of the engine, each consisting of a rubber mount for carrying the vertical loads and a torque arm for bearing the lateral forces (Mayer *et al.*, 2006). Thus, at each of the four mounting positions, the input admittance is measured in three directions. The transfer function from the exciting force to different positions inside the passenger compartment is measured as well. For the integration into a simulation, parametric models are fitted to the experimentally gained data as can be seen in Fig. 10.29. Different identification techniques in the frequency and in the time domain are tested, e.g., the Steiglitz–McBride algorithm that works with a synthesized impulse response. Furthermore, variations of model structure, order, and sampling frequency are examined in order to get suitable models. As a result, an IIR (infinite impulse response) filter of order 8 is used for modelling the admittance, while an FIR (finite impulse response) filter of order 254 fits the transfer functions to the microphone positions inside the car. The simulation model is time-discrete with a sampling rate of 2 kHz. The relatively low sampling rate allows for speeding up the computational time; therefore, it should be possible to implement some time consuming control concepts in parallel.

In a first test, the measured force at the mount is used for feedback. As an example, results for the application of a PID controller for the realization of an integral force feedback control are shown in Fig. 10.30. A significant reduction of the vibration level of the car body at the engine mount position as well as a reduction of the interior noise at the microphone position can be reached. A force of 40 N is necessary for the active hybrid mount. As a second control concept, an adaptive feed-forward controller is tested. The implemented FxLMS algorithm is quite common in active noise or vibration control applications when the disturbance source is known and a suitable reference signal can be used. As an error sensor signal is necessary for the adaptation of the control filter, the velocity at the mounting point is used. The results are also shown in Fig. 10.30. Obviously, the adaptive controller is more effective for higher frequencies, while the feedback controller reduces amplitudes at low frequencies.

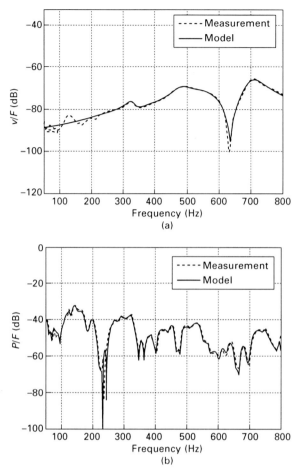

10.29 Measured point admittance of the car body at the mount position and fitted model (a); measured transfer function to the microphone at the driver's left ear and fitted model (b).

10.4.5 Active tuned vibration absorber

All modern cars are nowadays equipped with HVAC (heating, ventilation, and air conditioning) units. These units normally consist of heat exchangers, a compressor, a condenser fan, and a ventilation fan and can become rather large (Fig. 10.31) for commercial vehicles or rail vehicles. These components cause vibrations and noise, often at an annoying level, both on the inside and on the outside of the vehicle. For such a large HVAC unit, typical of those used in trams, active tuned vibration absorbers were designed that reduce the vibrations caused by the compressor (Bartel, 2008).

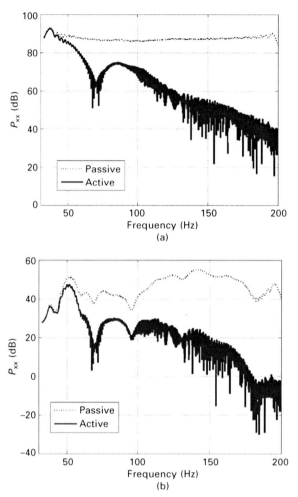

10.30 Results of control simulations: velocity at the mounting point (a) and sound pressure at the driver's left ear position (b).

The design of the active tuned vibration absorber was inspired by a paper by Konstanzer *et al.* (2006). Figure 10.32 shows a schematic view of the active tuned vibration absorbers mounted beneath the compressor. The vibration absorbers consist of two discrete masses attached to the ends of two cantilevered beams and are tuned to 50 Hz, which is very close to the frequency of the highest vibration level of the compressor. This fundamental frequency can be varied within a certain range by an acceleration feedback control system using the acceleration of the discrete masses as input to the voltage signal applied to the piezoelectric patch actuators attached to the cantilevered beams, thus virtually adapting the mass of the passive absorber. In addition, the acceleration at the mounting point can be used as the input

10.31 General view of the HVAC unit.

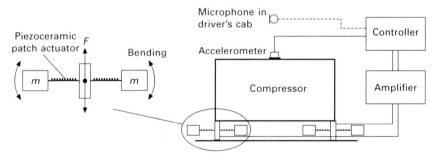

10.32 Schematic view of the active tuned vibration absorbers
mounted beneath the compressor.

to another control system such that the absorber behaves as a vibration
compensator at higher frequencies, using the inertia of the masses at the
end of the cantilevered beams to generate a force at the mounting points.

Figure 10.33 shows the active tuned vibration absorbers mounted beneath
the compressor. The acceleration of only one of the two masses of each
absorber was fed back as a voltage signal to both piezo patches instead of
applying an individual voltage signal to each of the two patches. The six
discrete masses of the three vibration absorbers have a total mass of 3.5 kg,
which is approximately 10% of the compressor's mass. The acceleration levels
of the HVAC unit's housing directly beneath the compressor are depicted
in Fig. 10.34. The dashed and solid lines show the acceleration levels with

10.33 Active tuned vibration absorbers mounted beneath the compressor.

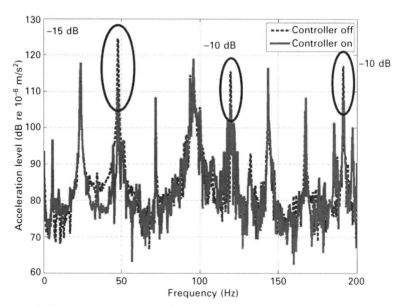

10.34 Significant reduction of the acceleration level beneath the compressor at 48 Hz and other frequencies.

the controller being switched off and on, respectively. Two effects can be seen:

1. At 48 Hz the acceleration levels are reduced by 15 dB by means of the passive vibration absorber whose fundamental frequency is actively tuned to match exactly the frequency of the highest peak.
2. At 119 Hz and 191 Hz the vibration absorbers act as vibration compensators, thus reducing the acceleration levels by 10 dB. At these frequencies an appropriate anti-phase sinusoidal voltage signal is applied to the piezo actuators. The required amplitudes and phase shifts are determined automatically by a digital control system. The peaks at 119 Hz and 191 Hz were chosen arbitrarily in order to demonstrate the potential of the compensator effect – other or more peaks could be reduced as well.

10.5   Conclusions

The drivers for developing advanced technologies for optimized NVH properties by means of passive and active systems can be summarized as:

1. to meet regulations imposed on the automotive industry for the control of emissions, fuel efficiency, and safety; and
2. to even gain a competitive edge by providing superior vehicle comfort in terms of noise and vibration characteristics.

These two drivers are very compelling and not necessarily independent of each other. The regulations, on the one hand, that drive the development of passive and active measures for NVH control for automotive markets originate from a worldwide concern for the environment and the issues of clean air and clean water for a burgeoning population that is unfortunately increasing its use of fossil fuels at an even greater rate than the population growth would suggest. On the other hand, the automotive market is one of the most regulated and probably more competitive than other markets.

The focus of today's R&D projects in the automotive industry lies on the improvement of lightweight design, product quality impression, comfort, and life cycle cost as well as noise emission, pollution, and safety, where the latter are particularly driven by legal requirements. This trend will continue while a special focus will be on exploiting advanced passive material systems as well as on the known multifunctional, intelligent material systems. For the latter it can be observed that most OEMs and Tier 1 suppliers – supported by R&D facilities – investigate the potential of smart structures for various applications. Most automotive R&D projects involving smart structures are concerned with AVC and ASAC for optimization of NVH characteristics and even sound design. However, the applied concepts for adaptive structural systems can be considered as a cross-cutting technology over all fields of today's engineering, but particularly for the transport sector.

For the automotive industry, a switch to advanced materials and active system for noise and vibration control would mean fewer parts and less weight. A 10% weight reduction translates into about 4% fuel and emission reduction, and that is why weight reduction ranks among the main targets of the transport industry to improve environment friendliness. This gain can be highly increased when a holistic approach for controlled NVH properties are proved viable and widely used in the vehicle structure. According to a very conservative estimation of the automotive industry, e.g., assuming an average yearly production of 350 000 cars over eight years and a conventional mileage of 150 000 km per car, a 10% weight reduction already gives a total saving of 61 000 tons of fuel, which means 200 000 tons less CO_2 released to the atmosphere. Considering, in addition, the possibility of these high performing material systems obtaining an extended component's life cycle or a reduction by 50% of the time-to-market for new material integration in the transport supply chain, these numbers altogether illustrate the real potential of environmental hazards savings.

10.6   Acknowledgements

Parts of the research presented in this chapter were performed in the framework of the integrated European project 'InMAR – Intelligent Materials for Active noise Reduction' (NMP2-CT-2003-501084) and the LOEWE-Zentrum AdRIA funded by the German federal state Hessen. The financial support by these projects is gratefully acknowledged. Other parts of the research presented here were performed in the framework of a Fraunhofer internal project. The authors are indebted to all who contributed to those projects. Furthermore, the authors wish to thank Dr Rainer Storm (TU Darmstadt), Finn Kryger Nielsen (Brüel & Kjaer), Andrew Christie (PPG) and Hans-Martin Gerhard (Porsche) for granting the use of lecture materials and pictures.

10.7   References

ATZ (2004), 'Adaptives Fahrwerk im Vectra', *ATZ* 09/2004, Jahrgang 106, 751.
Backfisch K P (2006), 'Hightech als Ladenhüter', *Automobil Industrie*, 06, 58–60.
Bartel T (2008), *Entwicklung eines adaptiven Schwingungsabsorbers für ein Straßenbahnklimagerät*, Diploma Thesis, Technische Universität Darmstadt, 2008
Bein Th *et al.* (2005), 'Adaptronik – ein technischer Ansatz zur Lösung bionischer Aufgaben', in Tropea C and Rossmann T (eds.), *Bionik*, Berlin, Springer-Verlag, 17–36.
Bein Th *et al.* (2008), 'Smart interfaces and semi-active vibration absorber for noise reduction in vehicle structures', *Aerospace Science and Technology*, 12, 62–73.
Blöcking F and Sugg B (2006), 'Piezo actuators: a technology prevails with injection valves for combustion engines', *Proceedings of the Actuator 2006*, 10th International Conference on New Actuators, 14–16 June, Bremen, Germany, 171–176.
Bös J and Mayer D (2006), 'Design and application of a semi-active electromechanical

vibration absorber', *Proceedings of the Thirteenth International Conference on Sound and Vibration (ICSV13)*, 2–6 July, Vienna, Austria.

Brennan M J and Kidner M (1995), 'Improving the performance of a vibration neutraliser by actively removing damping', *Journal of Sound and Vibration*, 221 (4), 587–606.

Christie A (2010), 'Developments in acrylic-based aqueous LASD driven by advances in polymer science', *Proceedings of the 6th International Styrian Noise, Vibration & Harshness Congress*, The European Automotive Noise Conference, 9–11 June, Graz, Austria.

Elliott S (2008), 'A review of active noise and vibration control in road vehicles', *ISVR Technical Memorandum No 981*, Southampton, UK.

Fursdon P M T, Harrison A J and Stoten D P (2000), 'The design and development of a self-tuning active engine mount', *European Conference on Vehicle Noise and Vibration*, IMechE, 21–32.

Gardonio P (2006), 'Arbitrary active constrained layer damping treatments on beams: finite element modelling and experimental validation', *Computers and Structures*, 84, 1384–1401.

Gardonio P and Brennan M J (2002), 'On the origins and development of mobility and impedance methods in structural dynamics', *Journal of Sound and Vibration*, 249 (3), 557–573.

Goroncy J (2005), 'Hier federt der Strom', *Automobil Industrie*, 1–2, 54–56.

Gruber P, Winner H, Hartel V and Holst M (2003), 'Beeinflussung des Fahrverhaltens durch adaptive Fahrwerklager', *Proceedings of VDI-Tagung Reifen-Fahrwerk-Fahrbahn*, VDI-Berichte 1791, 29–30 October, Hannover, Germany, 171–175.

Hagood N W and von Flotow A (1991), 'Damping of structural vibrations with piezoelectric materials and passive electrical networks', *Journal of Sound and Vibration*, 146 (2), 243–268.

Hanselka H and Nordmann R (2011), 'Kapitel O: Maschinendynamik', in *Dubbel – Taschenbuch für den Maschinenbau*, 23. Auflage, Springer Verlag 2011.

Herold S, Atzrodt H, Mayer D, Thomaier M and Melz T (2005), 'Gesamtsystemsimulation aktiver Strukturen am Beispiel eines aktiven Interfaces', *Proceedings of Internationales Forum Mechatronik*, 15–16 June, Augsburg, Germany.

H&H (2006), CoustiPlate, data sheet 2006, issue 2. Available from: http://www.acoustic.co.uk [Accessed 15 November 2010].

Hollkamp J J (1994), 'Multimodal passive vibration suppression with piezoelectric materials and resonant shunts', *Journal of Intelligent Material Systems and Structures*, 5, 49–57.

InMAR (2008) 'Publishable Executive Summary'. Available from www.inmar.info.

Janocha H (2006), 'Steuerbares Motorlager mit magnetorheologischer Flüssigkeit – Controllable engine mounting with MRF', AUTOREG 2006, *VDI-Berichte* Nr. 1931, 313–326.

Kalinke P and Gnauert U (2002), 'ATC: active torsion control zur Optimierung des Schwingungskomforts bei Cabriolets', *Proceedings of the Adaptronic Congress 2002*, 23–24 April, Potsdam, Germany.

Kalinke P, Gnauert U and Fehren, H (2001), 'Einsatz eines aktiven Schwingungsreduktionssystems zur Verbesserung des Schwingungskomforts bei Cabriolets', *Proceedings of the Adaptronic Congress 2001*, 4–5 April, Berlin, Germany

Konstanzer P et al. (2006), 'Piezo tunable vibration absorber system for aircraft interior noise reduction', *Proceedings of Euronoise 2006*, 30 May–1 June, Tampere, Finland.

Krix P (2006), 'Mehr als nur Schmuckstück – der Audi TT wird erwachsen', *Automobilwoche edition 2006*, 20–21.

Kumar N (2009), 'Vibration and damping characteristics of beams with active constrained layer treatments under parametric variations', *Materials and Design*, 30, 4162–4174.

Marienfeld P M, Bohn C, Karkosch H-J and Svaricek F (2002), 'Reduzierung des motorseitig eingeleiteten Körperschalls durch Einsatz adaptiver und aktiver Lagersysteme', *Proceedings of Global Chassis Control*, Haus der Technik, 3–4 December, Essen, Germany.

Matsuoka H, Mikasa T, Nemoto H (2004), 'NV countermeasure technology for a cylinder-on-demand engine mount', *Proceedings of SAE World Congress* (Paper 2004-01-0423), 8–11 March, Detroit, USA.

Matthias M, Thomaier M and Melz T (2005), 'Entwicklung, Bau und Test eines multiaxialen, modularen Inferfaces zur aktiven *Schwingungsreduktion für automotive Anwendungen*', *Proceedings of Adaptronic Congress*, 3–4 May, Göttingen, Germany.

Mayer D *et al.* (2006), 'Modeling of an active engine mount system for automotive applications', *Proceedings of EuroNoise 2006*, 30 May–1 June, Tampere, Finland.

Mayer D, Atzrodt H and Melz T (2009), 'Reduction of bearing vibrations with shunt damping', *Proceedings of Sixteenth International Congress of Sound and Vibrations*, 5–9 July, Krakow, Poland.

Mayer D, Herold S, Kauba M and Koch T (2010), 'Approaches for distributed active and passive vibration compensation', *Proceedings of ISMA 2010*, 20–22 September, Leuven, Belgium.

Melz T (2001), *Entwicklung und Qualifikation modularer Satellitensysteme zur adaptiven Vibrationskompensation an mechanischen Kryokühlern*, Dissertation, Darmstadt.

Melz T and Matthias M (2005), 'The Fraunhofer MaVo FASPAS for smart system design for automotive and machine tool engineering', *Proceedings of the 12th SPIE International Symposium*, 6–10 March, San Diego, California, USA.

Melz T, Mayer D and Thomaier M (2007), 'Adaptronic systems in automobiles', in Janocha H (ed.), *Adaptronics and Smart Structures: Basics, Materials, Design and Applications*, 2nd revised ed., Springer-Verlag.

MSC (2010), 'Engineered Quiet'. Available from: http://www.quietsteel-europe.com [Accessed 15 November 2010].

Sano H, Inoue T, Takahashi A, Terai K and Nakamura Y (2001), 'Active control system for low-frequency road noise combined with an audio system', *Proceedings of IEEE Trans. Speech and Audio Processing 9*, 755–763.

Schmidt K, Thörmann V, Weyer T, Mayer D, Herold S and Krajenski V (2003), 'Aktive Schwingungskompensation an einer PKW-Dachstruktur', *Proceedings of Adaptronic Congress 2003*, 1–3 April, Wolfsburg, Germany.

Storm R and Hanselka H (2008), *Kompendium Maschinenakustik, Bd. 1&2*, lecture script, TU Darmstadt, Typographics GmbH, Darmstadt, Germany.

TKS (Thyssen Krupp Stahl) (2001), 'Bondal – composite material with structure-borne sound damping properties', Advertising March 2001 edition. Available from: http://www.thyssenkrupp-stahl-service-center.com [Accessed 15 November 2010].

Viana F A C and Steffen Jr V (2006), 'Multimodal vibration damping through piezoelectric patches and optimal resonant shunt circuits', *Journal of the Brazilian Society of Mechanical Sciences and Engineering*, 28, 293–310.

11
Recycling of materials in automotive engineering

K. KIRWAN and B. M. WOOD, WMG,
University of Warwick, UK

Abstract: Meeting the targets set out by the EU's End of Life Vehicle (ELV) Directive is a challenge currently facing the automotive industry. In order to meet the requirements of this legislation, vehicles must be designed, manufactured and assembled giving consideration to their disposal. The reuse, recycling and recovery of ELV materials each have their own advantages and disadvantages, and are complicated by the wide range of metals, polymers and composites currently used in the manufacture of vehicles. Another option is the design of multifunctional components to reduce the number of end of life materials that need to be dealt with. Tools such as life cycle assessment (LCA) are available to evaluate the sustainability of a vehicle across the whole life cycle.

Key words: automotive, end of life vehicle, recycling, reuse, recovery, sustainability, life cycle assessment.

11.1   End of life vehicles (ELVs)

Given the impact that automotive vehicles have on our everyday lives, the resources they consume and the volumes of waste that they generate, it is unsurprising that the industry has been subjected to an ever increasing amount of legislation targeted at minimising their detrimental attributes upon society. The EU's End of Life Vehicle (ELV) Directive presents some of the greatest challenges to the automotive industry in the 21st century, as it forces radical thinking about the methods by which vehicles are designed, built and ultimately disposed of. This chapter will discuss the practical impacts of the ELV directive, the issues that have arisen since its adoption into national legislation, look at how they are being addressed by interested stakeholders in the community and what future alternative materials could be employed to overcome some of today's issues.

11.1.1 Overview of ELV

European Union legislation concerned with end of live vehicle (ELV) disposal was brought into existence in October 2000, giving member states of the EU a deadline of 21st April 2002 to ensure individual complicity. Although

299

this deadline for national complicity was missed by several states, the UK initially adopted the regulations on 3rd November 2003, with subsequent extensions in 2005 (Producer Responsibility) and amendments scheduled for 2010. [1]

The legislation applies to all passenger vehicles with no more than eight seats in addition to the driver's seat and goods vehicles with a mass not exceeding 3.5 tonnes, and enforces a disposal process of such vehicles, namely:

- From 1st January 2007, vehicle owners can have their complete ELVs accepted by collection systems free of charge, even when they have a negative value.
- Producers (vehicle manufacturers or professional importers) must pay 'all or a significant part' of the costs of take back and treatment for complete ELVs.
- Rising targets for reuse, recycling and recovery (including energy recovery) had to be achieved by economic operators by January 2006 and by 2015. Article 7 required economic operators to attain a reuse and recovery target of 85% for all ELVs by 1 January 2006 and within this, a target of 80% for reuse and recycling, increasing to 95% and 85% respectively in 2015.

In essence this piece of legislation saw a paradigm shift throughout the automotive world, where disposal of vehicles could only be carried out at an authorised treatment facility (ATF). The role of these facilities is to 'depollute' the vehicles so that all toxic and undesirable materials and fluids are removed to render the ELV as non-hazardous waste, and direct constituent parts of said vehicle into suitable reuse and recycling streams. These facilities are also responsible for documenting and evidencing this activity so that the UK government can demonstrate compliance with the ELV directive.

More fundamentally, the principle of producer responsibility means that they should increasingly take responsibility for their products once they become waste and ensure that vehicles are increasingly designed for recycling, limiting the usage of hazardous substances and utilising greater quantities of recycled materials in their new vehicles. This is reinforced by the fact that vehicle manufacturers have been directly responsible for any costs associated with ELV disposal of one of their product range sold after 1st July 2002 – the easier their vehicles can be recycled, the cheaper it will be.

11.1.2 The problems with ELVs

As with all legislation, the ELV directive has not been without problems in implementation and maintenance. Vehicles are, by their very nature, complex things and the sheer numbers in circulation mean that the practical logistics of disposal are immense. When the huge variation in constituent materials, deconstruction requirements, downstream processes and even vehicle ages are taken into consideration, it is perhaps amazing that the directive has been practically implemented in any way. It is widely accepted that approximately 75% of a vehicle was already recycled before the ELV directive came into force [2], simply driven by the scrap metal value inherent within the vehicle. A breakdown of a typical car is shown in Fig. 11.1. [3]

The challenge to the automotive industry, both manufacturers and disposers is how to deal with the remaining 25% of the vehicle that remains stubbornly problematic. The most obvious focus to achieve the 85% requirement is plastics which make up the largest non-metallic part of a vehicle, and with an ongoing drive towards lightweight vehicles for improved efficiency are likely to increase in terms of vehicle fraction. However, whilst recycling plastics is often perceived to be a simple exercise, the reality is very different and will be discussed later in this chapter.

The numbers involved

There are approximately 26 million vehicles currently in the UK and around 2 million end-of-life vehicles (ELVs) arise in the UK each year. There are two broad categories of ELVs; relatively new cars which result from accident

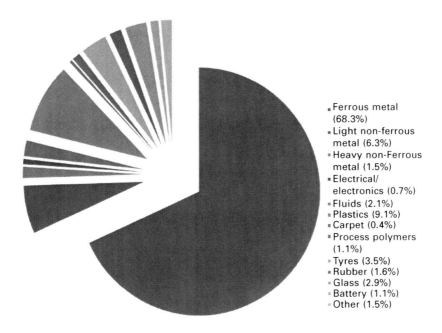

Ferrous metal (68.3%)
Light non-ferrous metal (6.3%)
Heavy non-Ferrous metal (1.5%)
Electrical/ electronics (0.7%)
Fluids (2.1%)
Plastics (9.1%)
Carpet (0.4%)
Process polymers (1.1%)
Tyres (3.5%)
Rubber (1.6%)
Glass (2.9%)
Battery (1.1%)
Other (1.5%)

11.1 The typical breakdown of materials in an ELV.

write-offs, known as premature ELVs and cars which have reached the end of their life naturally or natural ELVs. Natural ELVs often arise due to MOT failures and have an average life of around 12 to 13 years. [3]

11.1.3 The ELV process

When a vehicle reaches the end of its useful life, it will be taken to an authorised treatment facility (ATF) which has to accept it free of charge from the last owner. In addition to accepting vehicles, ATFs within the UK also have to collect ELVs free of charge from anyone who lives more than 30 miles away from such a facility. There are estimated [3] to be between 2,000 and 3,500 companies of differing sizes within the UK operating as ATFs although this number does appear to fluctuate dramatically over time.

When a vehicle enters an ATF, it will pass through a number of different stages aimed at breaking the vehicle into its constituent material or component parts for subsequent recycling or reuse. A process flow diagram for ELV disposal is shown in Fig. 10.2. Initially it will be depolluted. During this activity the battery is removed, followed by the tyres, fuel tank and any explosive devices such as air bags/seatbelt pre-tensioners (which can also be detonated *in-situ* with controlled explosions). All of the fluids within the vehicle such as fuel, motor oil, transmission oil etc. are then drained (often through drilled holes and/or suction from the vehicle) and stored for subsequent recycling/treatment. Other hazardous materials such as lead

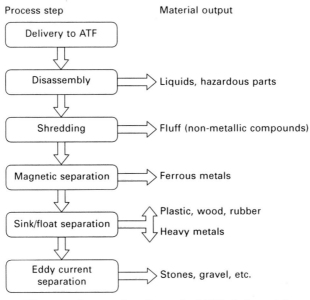

11.2 Process diagram for disposal of ELVs (adapted from [4]).

balancing weights and mercury switches are also removed at this stage and sent for specialist reprocessing.

Once the vehicle has been depolluted, it is then most likely to be crushed on-site and moved to a large shredder facility (of which there are only 37 in the UK), where all of the 'heavy' materials such as ferrous metals, copper, brass and aluminium are recovered. The 'light' materials including plastics, carpets, foams, etc., are now left and would traditionally have gone to landfill. However, the legislative requirements for 85% reuse or recycling means this can no longer occur, although incineration with subsequent energy recovery remains an alternative option for some of these materials.

Whilst a considerable amount of research effort is going into finding methods of recycling and reusing this light fraction, there are few, if any success stories to report. Even from this brief overview of activities, it can be seen that meeting the legislative requirements has been (and still is) a major challenge commercially, technically and logistically.

11.2   Reuse, recycle or recover?

In order to meet these challenging requirements there are a number of treatment routes that can be used to dispose of materials from ELVs. The most widely known are reuse, recycling and recovery but the use of natural methods of disposal such as biodegradation can also be implemented. It is also worth considering how we can reduce the amount of materials that we need to use in automotive products by investigating the use of novel materials that either allow a reduction in mass or an increase in strength.

11.2.1 Reuse

The lowest environmental impact results from a material or component which can be reused directly at the end of its usable life with little or no further processing. This is unfortunately very difficult to achieve in an automotive context, hence there are a number of options for end of life treatment that either process used materials such that they can be used as raw materials in a new process, or to recover some of the energy lost during their manufacture.

Items such as glass milk bottles can be cleaned, refilled and reused almost indefinitely, minimising the input of raw materials and energy and minimising packaging waste. In the automotive industry and particularly with plastic materials this process is not generally possible due to the huge variety and complexity of parts that exist.

Reuse is the end of life solution with the minimum environmental impact, as it extends the lifecycle of a material or component and therefore reduces the requirement for more raw materials or process energy.

An example of successful reuse of automotive products can be seen in the developing world where engines and other components from ELVs have been used to make water pumps for areas where clean water is difficult to obtain. By finding another application for an existing component from the automotive industry and using simple technology to adapt it for further use without the need to break it down into individual materials for reprocessing, this represents a significant saving of cost and energy compared to any of the other end of life processing techniques.

11.2.2 Recycling

Recycling a material involves recovering an end of life component, reducing it to a usable raw material and then using this to manufacture a new component. The benefits of recycling include reduction in the consumption of resources, reduced waste going to landfill and lower embodied energy.

Metals such as steel and aluminium are relatively easy to recycle, and in the case of aluminium this reduces the greenhouse gas emissions resulting from production by 95%. [5] This is because the temperature required to melt the recycled material (660 °C) is lower than the melting point of the aluminium ore (900 °C). Virgin aluminium is smelted using the Hall–Héroult process [6] which is highly energy intensive. Liquid aluminium is obtained from alumina by an oxidation reaction and is deposited electrochemically at an anode. The anodes used in this reaction are made from carbon and are therefore costly, particularly as they are consumed during the reaction. The reaction also results in the production of carbon dioxide. [7]

Steel recycling typically takes place in an electric arc furnace, although some oxygen blast furnaces are used to remelt scrap steel. When recycled steel is used in place of steel from primary sources, it has been calculated that a CO_2 saving of 1.8 tonnes (or 80%) is realistically achievable. [8]

It is also possible to recycle thermoplastic polymers such as polypropylene (PP) by remelting them and using them to manufacture new components. This process is not indefinite, however, and there is a limit to the number of times a thermoplastic can be recycled before the properties degrade. [9] In order to recycle plastic waste the individual polymers must be separated before they are remelted for further processing – this presents the biggest challenge for the efficient recycling of plastics. Given the large variety of polymers that can be found within a typical ELV, it explains why the desirable 'plastics' portion in Fig. 11.1 remains stubbornly illusive to tackle. Efficient recycling of polymers can be further complicated by contamination from ELVs due to the number of other materials present within the waste from automotive vehicles. The proper separation, segregation and treatment of polymer materials from ELVs therefore remains challenging, but essential if more of the vehicle is to be recycled rather than incinerated for energy recovery. One way of tackling the plastics issue is to develop a 'closed loop' recycling model, which if combined with a reduction of polymer types utilised within a vehicle, could make some vital contributions towards achieving ELV requirements.

In the electronics industry, Sony have attempted to do this by offering to recycle their products free of charge when they are finished with. [10] Because their products have been specifically designed with end of life disposal in mind, Sony know what materials are present in each model and it is possible to efficiently retrieve materials from end of life products, recycle the materials and then incorporate them into new products. Aside from obvious environmental benefits and legal compliance (in this instance with the WEEE directive), their reliance and financial outlay on virgin material is reduced. This model of recycling (or a workable variant) has yet to be adopted by the automotive industry, but the benefits could bring similar reductions in the cost and environmental impact of the raw materials used for the manufacture of automotive vehicles, albeit over a prolonged time compared to the electronics industry.

Composite materials are more difficult to recycle because they have two or more phases which are combined to make one material. The strong bond between these phases helps to embue composites with the excellent mechanical properties which make them so attractive, and separating these phases for recycling can therefore be challenging. The majority of end of life composites are disposed of by landfilling.

11.2.3 Energy recovery

Some of the embodied energy of end of life components can be recovered if they are burned and the heat generated is used to generate electricity. Energy from waste plants which turn municipal solid waste (MSW) into electricity are used across the UK to dispose of general waste which would otherwise go to landfill. This waste treatment route is a possible solution for the disposal of the non-metallic components of end of life vehicles.

Although recycling of materials such as aluminium and glass has been shown to considerably reduce process energy when making a new product, the recycling of plastics is not as energy efficient because there is little difference in the energy needed to manufacture a part from recycled or virgin polymer. However, plastics are generally manufactured from petrochemical feedstocks and therefore retain a large proportion (up to 99%) of the energy content of the crude oil derivatives that they are made from. The combustion of waste plastics allows a large proportion of their embodied energy to be usefully recovered [11] which can be used to subsidise or replace oil in the energy generation sector.

It has been suggested that the combustion of plastics alongside MSW can result in an increase in harmful emissions to atmosphere; however, in a comprehensive review of the role of plastics in energy recovery Mirza [12] found that the addition of plastics had no effect on the concentrations of heavy metals and trace organics in air emissions or solid residues.

A problem for energy recovery of automotive waste is the combustion of glass filled polymers. These materials use short glass fibres to reinforce a polymer component, but this can severely reduce incinerator efficiency and therefore limit the total energy that can be recovered through this process. Automotive manufacturers such as Mercedes-Benz [13] are adopting natural fibres to replace glass as fillers for thermoplastic polymers due to their low embodied energy, renewability and the ease with which they can be burnt at end of life to recover energy.

Energy recovery is an attractive solution if it is not possible to reuse or recycle a component as it offsets some of the energy embodied within a material. The negative aspects include the production of harmful gases from the combustion process, and the labour and energy input needed to run the process.

11.2.4 Biodegradation

Biodegradation is the chemical breakdown of materials by the environment. It is usually only relevant to biomass which is actively consumed by microbial or enzymatic digestion (in a manner similar to composting); however, some synthetic materials such as green plastics are similar enough to natural materials to be broken down in this way [14, 15]. The main benefit of biodegradable materials is the potential reduction of waste compared to synthetic polymers which can remain in the environment for hundreds or thousands of years if put to landfill. The breakdown of materials by microbes and enzymatic digestion can, however, result in the production of methane and other gases which have been associated with the 'greenhouse effect'. Biodegradation does not necessarily occur for all materials by simply leaving them exposed to the elements, and industrial composting may be used to facilitate the breakdown of biodegradable materials in a controlled environment, although there are a limited number of facilities capable of doing this in the UK. [16]

If more biodegradable materials were utilised in the manufacture of cars this process would be of great interest to automotive manufacturers, requiring little or no process energy at end of life and producing little waste after degradation has occurred. The facilities to carry out effective biodegradation of materials suitable for automotive use, however, are currently limited and if poorly controlled the natural breakdown of organic matter can result in the production of greenhouse gases.

11.2.5 What about reduce?

Reducing the amount of materials used in manufacturing a component obviously reduces the amount and variation of material which has to be dealt with at end of life. This can be achieved by integrating several functions into the same component through careful design, thus reducing the number of components on the vehicle and hence the amount of material required. The disadvantage of this approach is the inherent increase in component complexity which it requires, thus causing a potential increase in the difficulty of their disposal at end of life and the energy and cost required in their initial manufacture. An example of this is the rear quarter window used on the Smart ForTwo car. This polycarbonate component has integrated seals, decoration and fastening clips (see Fig. 11.3), minimising assembly time and reducing the amount and variation of material required for manufacture.

11.2.6 The emergence of carbon fibre

Modern road cars, particularly those of a premium or high-performance nature, are making use of more advanced and exotic materials in an attempt to reduce weight and therefore increase their efficiency and performance without the need to remove equipment. Composite materials such as carbon fibre reinforced polymer (CFRP), previously only used in the most expensive supercars, are finding their way into more mainstream vehicles as the manufacturing technology and desire for weight reduction increases, especially in the emerging hybrids and electric vehicles where weight will be paramount. However, reusing or recycling carbon fibres at the end of

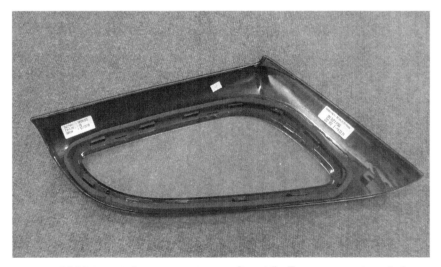

11.3 Integrated components on a Smart ForTwo rear quarter window.

life is challenging – the thermoset matrices cannot be remoulded, and the composite has a complex composition and is frequently contaminated. In the past, carbon fibre composites have been disposed mainly by landfilling [17] although a significant amount of academic/industrial research has been undertaken over the last 15 years [17,18] to firstly extract the carbon fibres and secondly reuse them in new composites. The commercial and technical viability of these approaches remains debatable and a large amount of research is ongoing in the area.

11.2.7 The influence of design on ELV costs

The emergence of ELV legislation presents a new challenge for automotive manufacturers. Figure 11.4 outlines the economic opportunity. The industry is generally cost driven and, in essence, the business model generally adopted is that of producing a car at lowest unit cost (model A) and until recently did not need to consider costs of disposal. However, with the costs and responsibility of disposal now falling clearly on the shoulders of manufacturers (model B); the lowest unit cost may not necessarily be the most viable option.

It is now quite conceivable that a more expensive initial design of vehicle may actually be more economically viable when disposal costs are taken into consideration (model C). Companies that are able to dispose of their vehicles more cheaply through clever design changes are actually able to make more profit (or less loss) than their competitors in the longer term if full life cycle costs are considered.

11.3   Environmental impact assessment tools

As environmental concerns have become more important to companies a variety of tools have been produced to help develop the sustainability of their products. One of the most popular is life cycle assessment (LCA), which considers the full life of a product from raw materials through to end of life or 'cradle to grave'. [19] The impacts and outputs of every stage of a product's life from raw materials to manufacturing to end of life are

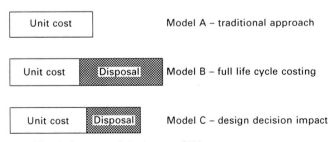

11.4 The influence of design on ELV costs.

calculated and logged. Typically an LCA study consists of the analysis of six life cycle stages: [20]

1. extraction of materials
2. manufacturing (including any refining processes, etc.)
3. packaging
4. transport
5. product use
6. end of life disposal (including recycling, recovery and reuse).

In the context of the automotive industry, LCA can be used to ensure that the materials used for manufacturing a vehicle will not result in environmental and disposal problems at end of life. An example of how these tools can help automotive manufacturers increase the sustainability of their products is in the use of lightweight materials such as magnesium [21], aluminium [22], and composite sandwich structures [23].

Tharumarajah and Koltun's study on the use of magnesium for lightweighting of automotive engine blocks found that there would be an advantageous reduction in energy usage and greenhouse gas emissions compared to cast iron or aluminium blocks, but the overall environmental benefits were dependent on the source of the magnesium used. [21] Das carried out a study on the life cycle impacts of replacing a steel boot panel with an aluminium one. The study showed that although the reduction in weight of the panel was beneficial in reducing the fuel consumption and therefore the energy used in the product use phase of the lifecycle, the higher embodied energy of the aluminium resulted in an energy payback time of 13 years. [22] Ermolaeva *et al.* included an environmental impact assessment into the materials selection and structural optimisation process of sandwich core structures for an automotive floor panel. [23] If viewed separately, the structural and environmental impact assessments resulted in different material recommendations; however, by combining the two processes a compromise was found which satisfied the structural requirements of the component while minimising environmental damage.

Although initially intended to evaluate the environmental aspects of a product's life, LCA can also include other parameters in order to ascertain the feasibility of a new product concept. A case study by Duval and MacLean of the potential of replacing virgin polymers for automotive panels with recycled material also included costings of the new and existing approaches to evaluate the project's economic as well as environmental sustainability. [24] Their case study concluded that although recycling end of life automotive plastics would result in energy savings and greenhouse gas reductions of approximately 50%, the resulting increase in costs would not be financially viable for the company in question.

Although useful tools for automotive companies, the results of LCA and other environmental impact assessments must be used with caution. Any study of this kind is based on a set of unique assumptions and therefore cannot be directly compared to another unless the assumptions used are identical. They are useful as an internal tool to evaluate new product concepts before going into the expensive new product development and prototyping process, and will allow automotive manufacturers to minimise the risk of future end of life disposal legislation by integrating sustainable thinking into the design and materials selection process.

11.4   Case study – the WorldF3rst racing car

The WorldF3rst racing car, a Formula Three car built by a team from WMG at the University of Warwick in the UK, provides an example of the use of a variety of novel and sustainable materials with benefits at end of life. By demonstrating these materials in the motorsport industry their effectiveness and suitability for use in a harsh environment has been proven. The car itself is shown in Fig. 11.5, performing the hill climb at Goodwood Festival of Speed in 2009.

The engine used in the WorldF3rst car is a two litre, turbocharged, four cylinder diesel engine taken from an end of life passenger car. The fuel used is a biodiesel manufactured from cocoa butter which has in turn been extracted from chocolate factory waste. The lubricating oil in the engine is made from plant oils and is therefore biodegradable in the right conditions. The engine produces approximately 230bhp and 420Nm of torque, giving it performance equivalent to a petrol powered F3 car.

11.5 WorldF3rst F3 Car tackling the hill climb at the Goodwood Festival of Speed 2009.

Body panels such as the engine cover, damper hatch and sidepods on the car are manufactured from a recycled carbon fibre which is reinforced with a polyester resin containing 30% recycled PET bottles. The carbon fibre is recovered from rolls of waste prepreg by pyrolysis, and then reimpregnated with the polyester resin before curing. The properties of the recycled PET resin were evaluated in previous work. [25]

The steering wheel is made from a carrot fibre reinforced polymer composite, which has mechanical properties between those of glass and carbon fibre composites. The carrot fibres used are a waste product derived from the pulp left after the manufacture of carrot juice.

Other natural fibre reinforced composite components include aerodynamic parts such as the bib which is made from a flax prepreg, and the barge boards which are made from a 3D woven hemp fabric. Even the wiring loom of the car was subjected to scrutiny; aluminium rather than copper was used for the wires, requiring a larger cross section to transmit the same current but reducing the final weight of the car. The polymer insulation used contained very low levels of harmful phosphorous, and all plastic connectors were designed for disassembly by colour coding them by material type. The approach used in the WorldF3rst car considers not only the sustainability of the materials used in its manufacture, but also the impact of their disposal at end of life. A breakdown of the materials used and their sources is given in Table 11.1. The end of life treatment routes for various components on the car can be seen in Fig. 11.6.

11.5   Conclusions

Meeting the targets set out by the EU's End of Life Vehicle (ELV) Directive is a challenge currently facing the automotive industry. In order to meet the requirements of this legislation, vehicles must be designed, manufactured and assembled giving consideration to their disposal when their useful life has ended. In practice, approximately 75% of an ELV has been readily recyclable for a number of years, the remaining 25% has proved frustratingly stubborn. The limits on landfill and energy recovery are challenging automotive engineers across the globe.

In an ideal world, the simplest and most environmentally friendly approach to ELV would be the reuse of components either within an automotive context

Table 11.1 Material Sources for the WorldF3rst F3 Car

Reused	Recycled	Renewable materials
Engine	Sidepods	Engine oil
	Engine cover	Steering wheel
	Damper hatch	Bib
		Barge boards

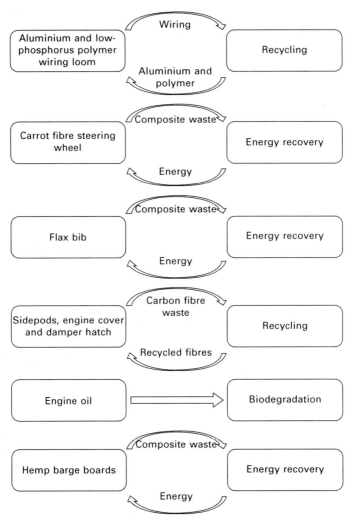

11.6 End of life treatment routes for the WorldF3rst F3 car.

or elsewhere. However, given the complexity of modern automotive systems and global legislation, reuse becomes in many instances impossible.

The physical recycling of ELV materials presents a different set of challenges. Metals are readily recyclable and there are a number of recognised environmental benefits. However, metals already represent the 75% of a vehicle that can be dealt with. Plastics are the next largest proportion of an ELV and they represent a much greater challenge. The differing range of plastics present within a vehicle, combined with the presence of contaminants and other materials means that it is often not practical or possible to recycle them back to usable materials. Energy recovery is a valid option for polymers, although the presence of reinforcements such as glass, minerals and carbon can complicate the process and reduce the efficiency of any incineration process.

Alternative approaches such as closed loop recycling and biodegradation have been used in other industries and are worth consideration in an automotive context. These approaches are very much in their infancy and limited in practical application. The environmental credentials of these approaches are uncertain and, in the case of biodegradation, could have a negative environmental impact if carried out incorrectly.

The final option for engineers at the present is a reduction in parts and/ or materials. Careful design of multifunctional components or integration of multiple parts into individual systems has the potential to reduce the number and variation of end of life materials that need to be dealt with. This approach has the potential to make end of life treatment cheaper and easier, but may very well increase initial component costs. The acid test will always be whether any initial increase in a component or system cost is offset at the end of life and will require a fundamental shift in automotive practices, moving away from the traditional 'lower unit cost' approach to a full life costing concept. In order to achieve this, environmental impact assessment tools such as LCA are available to designers to evaluate the potential impact of material choices before the design stage, and therefore ensure the sustainability of a vehicle across the whole lifecycle is considered.

11.6   References

[1] BIS, Department for Business Innovation and Skills, *The End of Live Vehicles Regulations 2003, 2005 and 2010, Government Guidance Notes*, June 2010
[2] Automotive Consortium on Recycling and Disposal (ACORD), *Annual Report*, 2001
[3] BERR, End of Life Vehicle (ELV) Waste Arisings and Recycling Rates, available online at: http://www.berr.gov.uk/files/file30652.pdf
[4] Cui, J. & Roven, H.J., Recycling of automotive aluminium, *Transactions of Nonferrous Metals Society of China*, 20 (2010) 2057–2063
[5] Global Aluminium Recycling: A Cornerstone of Sustainable Development, International Aluminium Institute, (2009), available online at: http://www.world-aluminium.org/cache/fl0000181.pdf
[6] Grjotheim, U. & Kvande, H., *Introduction to Aluminium Electrolysis: Understanding the Hall–Heroult Process*, Aluminium Verlag GmbH, (Germany), 1993, p. 260
[7] Rapp, B., Electrowinning aluminium, *Materials Today*, 9 (12) (2006) 6
[8] Avery, N., Coleman, N., Life cycle assessment methodologies for quantifying the benefits of steel reuse and recycling, *Construction Information Quarterly*, 11 (3) (September 2009) 127–131
[9] Goodship, V., Plastic recycling, *Science Progress*, (2007), 90(4), 245–268
[10] Sony, The Sony Take Back Recycling Program, Sony Electronics Inc. (2010), available online at: http://green.sel.sony.com/pages/recycle-2.html

[11] Lea, W.R, Plastic incineration versus recycling: a comparison of energy and landfill cost savings, *Journal of Hazardous Materials*, 47 (1996) 295–302

[12] Mirza, R., A review of the role of plastics in energy recovery, *Chemosphere*, 38 (1) (1998) 207–231

[13] Mercedes-Benz, Lifecycle – A Holistic Approach, available online at: http://www2.mercedes-benz.co.uk/content/unitedkingdom/mpc/mpc_unitedkingdom_website/en/home_mpc/passengercars/home/corporate_sales0/Fleet_manager_new/Our_cars/Environment/Lifecycle.html

[14] Stevens, E.S., Green Plastics: *An Introduction to the New Science of Biodegradable Plastics*, Princeton University Press, USA, (2002)

[15] Plackett, D., Biodegradable polymer composites from natural fibres, in Smith, R. (editor), *Biodegradable polymers for industrial applications*, Woodhead Publishing Limited, UK, (2005) 189–218

[16] Slater, R.A. & Frederickson, J., Composting municipal waste in the UK: some lessons from Europe, Resources, *Conservation and Recycling*, 32 (2001) 359–374

[17] Pickering, S. J. *Composites Part A*, 37 (2006) 1206

[18] Pimenta, S. & Pinho, S.T., Recycling carbon fibre reinforced polymers for structural applications: Technology review and market outlook, *Waste Management*, 31 (2011) 378–392

[19] Ljungberg, L.Y., Materials selection and design for development of sustainable products, *Materials and Design*, 28 (2007) 466–479

[20] Hui, E.K., Lau, H.C.W., Chan, H.S. & Lee, K.T., An environmental impact scoring system for manufactured products, *International Journal of Advanced Manufacturing Technology*, 19 (2002) 302–312

[21] Tharumarajah, A. & Koltun, P., Is there an environmental advantage of using magnesium components for light-weighting cars?, *Journal of Cleaner Production*, 15 (2007) 1007–1013

[22] Das, S., Lifecycle energy impacts of automotive liftgate inner, Resources, *Conservation and Recycling*, 43 (2005) 375–390

[23] Ermolaeva, N.S., Castro, M.B.G. & Kandachar, P.V., Materials selection for an automotive structure by using structural optimization with environmental impact assessment, *Materials and Design*, 25 (2004) 689–698

[24] Duval, D. & MacLean, H., The role of product information in automotive plastics recycling: a financial and life cycle assessment, *Journal of Cleaner Production*, 15 (2007) 1158–1168

[25] Wood, B.M., Coles, S.R., Kirwan, K. & Maggs, S., Biocomposites: Evaluating the Potential Compatibility of Natural Fibers and Resins for New Applications, *Journal of Advanced Materials*, 42 (2) (2010) 5–16

12
Joining technologies for automotive components

F. M. DE WIT and J.A. POULIS, Technical University
Delft, The Netherlands

Abstract: This chapter reviews key joining methods for automotive
materials. It first summarises the main classes of material used in automotive
engineering. The chapter then goes on to discuss different joining
technologies such as laser beam welding and brazing, adhesive bonding and
mechanical fastening techniques such as folding and riveting. It also reviews
hybrid joining systems.

Key words: automotive engineering, laser beam welding, brazing, adhesive
bonding, mechanical fastening techniques, folding, riveting, hybrid joining
systems.

12.1   Introduction

A priority for automotive engineers is reducing the weight of a car since
this is directly related to fuel economy. Developments like the design of
the power source and train (e.g. smaller engines with similar to better
performance than their larger predecessors) contribute to lowering overall
weight. However, making structural parts lighter, while still meeting safety
requirements, remains critical to decreasing weight. Using stronger steels
to create thinner panels has reduced weight significantly. Improvements
in modelling have also made it easier to optimise component strength and
stiffness whilst minimising weight. Another trend is to use lighter materials
(e.g. composites) for components that are not critical to structural integrity.
In recent years there has been a greater focus on recycling, which has
placed greater emphasis on the most efficient use of materials as well as
those materials that can most easily be recycled. The challenge of weight
reduction is greater because, while structural parts have become lighter, the
increased use of electronics, new power systems (e.g. batteries and engines
in electric and hybrid vehicles) and comfort systems (e.g. air conditioning)
have added significantly to car weight.

The requirement for thinner, lighter components made from a range of
materials creates special problems when it comes to joining. In an ideal
world 'the best joining method is no joining at all'. This quote captures
a designer's frustration with the need to include joints which then create

315

potential areas of weakness in the automotive body. However, in practice, the complexity of the modern automobile means joints are often inevitable. In this chapter we focus on methods for joining structural components in cars. The chapter begins with a brief overview of materials used in automotive engineering together with their most important properties in connection with joining. It then goes on to discuss key joining methods such as laser welding/brazing, adhesive bonding and mechanical bonding. Specifically, the chapter discusses the adhesive bonding of hybrid materials as well as hybrid joining techniques both for low as well as high-volume production. Finally, the chapter discusses trends for the (near) future.

12.2   Types of advanced structural materials in cars

High strength steels and other ferrous materials are still mainly used in the body in white (BIW) structures of production cars. The BIW consists of the various components of the car body joined together (e.g. by welding), but before moving parts (e.g. doors), the motor, chassis sub-assemblies or trim (e.g. seats, electronics) have been added. Steels are still used because of their relatively low price, high strength and adequate properties for joining technologies such as welding. The BIW still comprises up to 70% of the overall car weight. Attempts to reduce this weight have focused partly on using alternative lightweight metals, but this involves both technical problems (e.g. in retaining the requisite strength) and increases in component cost (250–350% of the cost of the comparable steel part). An alternative is the use of stronger steels with improved forming properties, which can therefore be used in thinner gauges to reduce weight. Combining different sorts of steel with different gauges in tailor made blanks has revolutionised weight saving. Apart from the BIW, there has also been an increasing trend in recent years to use other materials for car components. These include lightweight metals such as aluminium and magnesium as well as fibre-reinforced polymer composites (FRPCs). All these changes have brought new challenges in joining techniques.

12.2.1 (Ultra) high strength steels ((U)HSS)

In recent years, mild grades of steel have gradually been replaced by higher strength steels especially in the higher load-bearing and crash-resistant parts of the BIW. Advances in steel processing technology during the last three decades have resulted in steels with new properties. The basis for this improvement is the greater understanding of phase transformations in steels and the precision with which they can now be predicted and tailored to achieve particular properties. In broad terms, it has been possible to develop steels which combine higher (yield) strength with higher deformation. The

ULSAB project, for example, has shown that, by using (U)HSS in parts of the BIW, the overall vehicle can be made safer, lighter and stiffer at a competitive price. There are now several categories of high strength steels, each with specific yield strength to elongation properties, as shown in Fig. 12.1. Because of the special heat treatments used and higher alloying element concentration, these steels have become harder to weld. Other options such as adhesive bonding have been used to meet the low heat input required and to take account of the material's stiffness.

12.2.2 Aluminium alloys

In Europe, BIWs made out of aluminium account for about 34k metric tonnes of aluminium annually and, given the cost of the material, are mainly used in the lower volume/higher value segment of the automotive market [1]. Although there are weight savings to be achieved by using this material, it is still a minor material in most BIW structures. There are three main priorities for aluminium alloy research in the automotive sector. The first is to increase the strength of aluminium alloys to be able to make thinner sheets. The second is to control their formability whilst the third is to improve the material's impact resistance. An AA5xxx series alloy is a standard material for car parts. It is used, for example, in the inside construction of doors

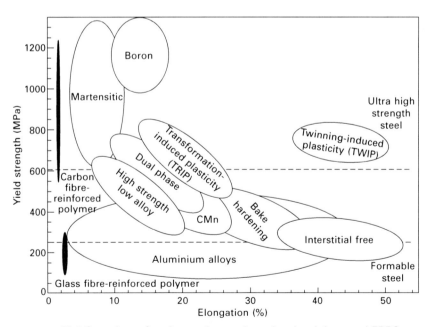

12.1 Overview of various advanced steels, aluminium and FRPCs used in production cars.

because of its excellent formability and corrosion properties. New material developments have shifted this application to recycled alloys as well as to AA6xxx alloys. For example, AA6082 is used in the Jaguar XJ 2010. It relies on processes like hydroforming because of the material's requirements. Recycled aluminium, for instance AA5754, is currently used in BIW structures. This adds to the total sustainability of the car and its cradle-to-grave life cycle, since aluminium materials are easily recycled (remelted) at only 5% of the energy compared to its primary production. It has been estimated that 516 million metric tonnes, roughly 75% of all aluminium produced since 1888 (709 million metric tonnes) is still in use today, and can theoretically be infinitely recycled. [1] As the heat sensitivity of these new alloys is much higher than steel, compounded by the use of thinner gauges, heat can spread more easily and alternative joining techniques such as laser welding and brazing need to be used.

12.2.3 Magnesium

Several parts of car bodies, for example, inside panel stiffeners, have been replaced by die-cast magnesium alloys. Although their corrosion resistance is rather low, the weight saving and their low melting temperature has traditionally given magnesium an advantage over aluminium alloys. They are, however, cumbersome to weld, but can be joined more easily with other joining processes like adhesive bonding or mechanical fastening.

12.2.4 Fibre-reinforced polymer (FRP) composites

In 1992 a carbon-fibre body was successfully used for the first time in the McLaren MP4/1 during Formula 1 races. Since then composites have been accepted as a viable construction material for the BIW but mainly for low volume/high price niche products [2], e.g. Lamborghini Avontador LP 700-4 or in concept vehicles (BMW MegaCity Vehicle and Alpha 4C). Fibre-reinforced polymer (FRP) composites are used more and more, specifically in small series vehicles and as part of the BIW where flat panels such as base plates are required. Joining of FRP composites can be achieved in the design of the joint, where inserts connect the parts and are (co)cured. Adhesive bonding and mechanical fastening are used as alternatives. Ultrasonic welding has also been used with thermoplastics to melt plastic locally and press parts together, effectively creating a new matrix where the polymers are mixed. However, the process is fairly sensitive to contamination and requires precise process parameter settings, i.e. surface temperature. Solvent bonding has been used with thermosets, but this is on the decline because of the ban on organic solvents.

12.3   Factors affecting the selection of joining methods

Selecting a certain joining technique depends on the materials that are needed, their properties and the surfaces to be joined. The strength and other properties required of the joint are also important factors. The joining technique chosen is also dependent on the costs associated with the method. This does not only mean the daily operating and maintenance costs, but also the investment costs necessary for training employees or purchasing equipment. The robustness and flexibility of the technique is another important factor. A highly efficient production line process relies heavily on the speed and effectiveness of the joining technique. Stopping a mass production line because of production problems is a considerable economic burden. In making a decision, the designer can make use of previous test results for particular joint types combined with, for example, finite element modelling (FEM) of the joint design under differing joining conditions. In real life, the selection of joining technology will be a compromise between technical, commercial and manufacturing factors.

Modern vehicle construction requires the use of different material technologies (hybrid materials and structures) and combinations of different joining technologies (hybrid joining). The most common material combinations for joining are:

- (U)HSS to (U)HSS
- one aluminium alloy to another aluminium alloy
- (U)HSS to aluminium alloy
- joining composite materials to either (U)HSS and aluminium.

There are several joining techniques which can be used to join hybrid materials (Table 12.1). Because of the increase in use of higher strength and lighter weight materials, traditional joining techniques face new challenges. Heat input is a much bigger problem than it used to be, since alloying and solidification processes have become much more critical. Another problem

Table 12.1 Overview of joining methods with respect to hybrid material combinations

Materials	LSW	RSW	(Laser)brazing	Adhesive bonding	Mech. fastening
UHSS/UHSS	Y	Y	N	Y	Y
Alu/Alu	Y	T	0	Y	Y
UHSS/Alu	N	N	Y	Y	Y
UHSS/FRP	T	N	N	Y	Y
Alu/FRP	T	N	N	Y	Y

(Y – possible and done; T – theoretically possible, but not applied in practice; N – not possible; 0 – technology not applicable)

is that for some materials, like FRP composites, very limited heat input is possible because the polymers involved exhibit a low melting point or, more importantly, a low T_g, below which their mechanical properties (e.g. the E modulus) can easily reduce to negligible values far below the expected values for effective use. As a result lower temperature (or lower heat input) techniques are often needed. These include:

- laser beam welding
- (laser) brazing
- adhesive bonding
- mechanical joining techniques like folding and riveting.

We will discuss these in later sections, focusing on hybrid material joints.

12.4   Joint design and joint surfaces

A joint can be seen as a transition from one material to the next and this transitional region has its own properties. The overall shape and requirements of the total structure will influence joint requirements and the methods of bonding. For instance changing from an internal framework that is then covered with a non-load-bearing skin to a unibody (using various materials) or monocoque (using one material – monocoque is French for 'single shell') design will require different joint designs.

The joint should be designed as one part of the overall construction. The stresses and strains on the materials to be joined and their behaviour under dynamic loading need to be quantifiable and predictable. If these are not known for the type of joint, a higher safety factor might have to be used. This would lead to over-designing the joint which, in turn, might lead to higher strength and potential failure in surrounding parts if stresses are not evenly distributed.

The configuration of joints is based on standard designs like overlap, butt, flanged and insert joints. Mechanical fasteners can also be used in certain configurations; these require special consideration, especially in allowing for dynamic loading. Each joining technique should be taken into account when designing a joint. Existing designs for one method of joining should generally not be carried through if a different joining technique is selected without a careful assessment and adaptation of the design. A change in materials would call for the same assessment. Any change will result in a new configuration of joining method, joint design and material. Some points which should be considered when changing from one joining technique to another are:

- The compatibility of heat input of the new joining method with the materials.

- The mechanical loading conditions, e.g. how much peel stress can the new joining method have?
- Does the shape of the joint lead to increased stress concentrations depending on the chosen joining technique?
- Is there another way of joining that can be added to enhance, for example, fatigue properties (adding adhesive bonding to resistance spot welding (RSW)), thus creating a hybrid joining technique.

For special cases and niche markets, the design can lead to creating a BIW without any distinguishable joints. The technique of integration of joints is already available for carbon fibre-reinforced polymer composite (CFRPC) materials. This kind of design process is, however, rather time-consuming and difficult, since it requires skilled manual lay-up and cutting of the CFRPC pre-preg materials. There are advances in the production techniques for these composites where two robots could be used to wrap the CFRPC materials around a foam or a detachable metal frame in a pre-planned sequence. A completely free-standing frame without any joints could therefore be created. In the future, speeding up this method could lead to as many as 4000 structures per annum, which would be impossible with existing manual methods [3].

12.4.1 Joint surfaces

Joining two materials is a matter of combining two free surfaces. The chemical state and composition of the surfaces determines in large part the quality of the final bond. Changing the surface can both affect the quality of a joint and even allow a different type of joining technique. Coatings for steels mostly use other less noble metals to protect against corrosion. The most widely used is zinc, which can be cladded to steels in various ways. Galvanising can be carried out by various processes involving depositing a thin layer of zinc on the substrate. In these zinc layers, other elements like magnesium might be present which potentially gives another surface composition and in case of adhesive bonding, other bonding characteristics.

In general, a lubricant is used to protect the metals from corrosion when in stock. This has to be removed before welding or adhesive bonding can take place. Alternatively, the lubricant is incorporated into the adhesive during the curing process. This is the case with modern 1K adhesives for automotive applications, which harden during the paint bake cycle. Contamination in general can lead to insufficiently strong joints if the contamination is incorporated into the base material or for instance the adhesive. Initially, the bond strength might not be so different from what would be expected but, after ageing by moisture exposure at elevated temperatures, such joints tend to fail at the interface due to the penetration of water onto the interface and subsequently corrosive attack on the base material.

12.5   Laser beam welding (LBW) and brazing/ soldering

Laser beam welding (LBW) can be considered a high-temperature technique. However, the main advantage of this type of joining is that the heat input can be fairly limited, because of the relatively small widths, which makes it ideal for metallic materials. It is not recommended to use LBW to make joints between (U)HSS and aluminium because of corrosion problems and the low level of mixing of one element into the other. Extremely brittle connections would be formed as well as creating problems with corrosion from the resulting galvanic coupling of the two materials. It is also not generally possible to use LBW with composite materials, as the melt surface temperature is around 1500°C. Fortunato *et al.* [4] have recently published a review of LBW of stainless steel to glass and carbon fibre reinforced PA66.

LBW generates a significant temperature differential between the molten metal and the base metal immediately adjacent to the weld. Heating and cooling rates are much higher in LBW than in arc welding, and the heat-affected zones (HAZs) are therefore much smaller. Because of the rapid cooling and relatively small HAZ, LBW leaves joints with low residual stresses, reducing the need for repairs. Rapid cooling rates can, however, create problems such as cracking in high carbon steels. It is possible to use LBW to join newer HSLA steels to twinning-induced plasticity (TWIP) or transformation-induced plasticity (TRIP) steels. LBW can also be used for joining aluminium alloys of different compositions. Because it is more cost-effective and flexible, LBW has become a relatively familiar and well-tried technique which has made it more popular than older welding techniques such as MIG welding. Disadvantages include the inability to do repairs after production. The technique is used regularly to make tailor made blanks, where pre-formed parts made from different sorts of steel are connected into one larger part.

12.5.1   (Laser) brazing/soldering

Diode lasers are now used mainly for brazing instead of Nd:YAG lasers. Advantages include the combining of non weld-compatible materials like aluminium and steels. Because of the low heat input in brazing, it is applicable to many different materials that are sensitive to heating once formed. The results of brazing are extremely clean looking and can therefore also be used on visible parts. The extra materials needed for brazing are, however, quite expensive, which limits its use. The technique is not suited for high load-bearing parts and for longer-term use since there are problems with fatigue.

12.6   Adhesive bonding

For structural purposes adhesive bonding offers many advantages. First of all, it is a technique that creates an increase in stiffness of the total construction if high modulus adhesives are used because of larger bond lines. The stress is therefore also distributed more evenly, which results in better fatigue properties. The standard used adhesive is a one-component (1K) epoxy. This epoxy has limited strength when not fully cured, and will only cure to its full strength in the paint bake oven. The initial strength just after the application of the adhesive, the so called 'green strength' is therefore not sufficient to guarantee the dimensional integrity of the bonded parts during the production process. For this reason adhesives are often combined with spot-welding or self-piercing rivets, which carry the loads during the production process until the adhesive is fully cured by the heat of the paint-bake oven.

Adhesive bonding can be an excellent way of connecting dissimilar materials because it does not require remelting or altering the mechanical properties of the materials being joined. Modern 1K epoxies do not require a surface pretreatment since they can incorporate oil residues and other contaminants. They can remove the need for sealants and add corrosion protection through electrical separation of otherwise galvanically-coupled metals. Inspection is possible by optical/visual techniques when the adhesive properties are known and the push-out of adhesive from the joint is minor. Because of their flexibility, the noise vibration harshness (NVH) levels in bonded components are low through the dampening of certain frequencies and vibrations. Adhesively bonded joints are repairable because of their ability to disbond through the application of sufficient heat input or forced shrinkage by rapid cooling to low temperatures (using, for example, liquid air). In metal structures they are compatible with melting processes in recycling, therefore not interfering with the recyclability of the whole structure.

Adhesive bonding can be easily combined with other joining techniques like traditional resistance spot welding (RSW), riveting and even LBW. Combined with mechanical joints, it is the only way to join fibre reinforced thermoset polymer composites, because it is compatible with this material's properties. Adhesives exhibit excellent fatigue properties in combination with high durability which, for example, enabled their early use in the aerospace industry when Dr Norman de Bruyne used phenol-formaldehyde resin toughened with polyvinylformaldehyde thermoplastic particles (still available in the market as Redux 775) in the Hornet Fighter in 1944.

Disadvantages of adhesive bonding include the relatively high quality of the substrates required. For high-strength properties precision in the proximity of the adherents is also crucial, since variations in thickness can affect the strength of the joint considerably. At best only small forces can be transferred when the adhesive has not (fully) cured. There are currently too few specific models simulating adhesively bonded connections which makes design more difficult. Adhesive bonding is not compatible with any form of significant heat input (e.g. from welding) once the adhesive has cured. Furthermore, large overlaps are necessary for high-strength material. The strength of adhesive bonds can be also be a problem if the materials being bonded are not themselves sufficiently strong. One example of this is joining multi-material stiffeners to too thin gauge plate materials which can then deform because of force exerted by the adhesive bonds.

12.7   Mechanical joints

There are various ways of joining materials using mechanical techniques. One of the oldest and most common techniques to fabricate car parts is folding, where malleable metals are wrapped around each other in various configurations. The advantage of this method is that it is robust and, within certain tolerances, creates a tight fit. Disadvantages are that only certain shapes can be made which can successfully incorporate folding and more material is needed because of the overlaps needed for the folds. Although folding results in a tight fit between the joined materials, sealing is sometimes still needed to prevent (crevice) corrosion due to the collection of dirt and water. From a production point of view, every part that is folded requires specific tooling, which adds to the costs of the technique and can be a constraint on higher volume production.

Riveting and bolting have traditionally been the main connection methods used in the aerospace industry, predominantly because of the robustness and ease of visual inspection for damage. Both riveting and bolting are available in many sizes and processes. An extensive overview is given by Messler [5]. Although this is a relatively cheap and fast technique, because of the need to drill holes with tight tolerances through more than one material, self-piercing rivets are generally used more in automotive manufacture. Another problem is disassembly because the rivets are usually made from different materials than the connected materials, leading to difficulties recycling the component. The fatigue properties associated with rivets are problematic because of the stress concentrations generated by the holes made in the parent materials. This is generally more problematic the stiffer and thinner the parent materials become. An additional problem is that in crashes, both bolted as well as riveted connections cannot take significant loading, since the rivets and bolts themselves are either ruptured or the materials in which they are embedded are torn.

12.8   Hybrid joining methods

The most frequently used hybrid joining techniques are based on adhesive bonding in combination with another structural joining process such as self-piercing rivets or spot welding. The main reason for these combinations is that when the adhesive is applied, it will have only a limited static strength until fully cured. Mechanical fastening provides additional strength until curing is complete. The longer term benefit of using adhesive bonding is that the stiffness of the overall final vehicle construction is considerably higher. Adhesive bonding also contributes to reducing the 'light weight index' which measures weight in relation to the overall mass of the vehicle.

12.8.1 Adhesive bonding/resistance spot welding (RSW)

If the strength of the green (uncured) adhesive is not sufficient to support the loads applied during subsequent processing, an additional hybrid joining technique is spot welding. This is only suitable for certain materials amenable to spot welding, for example, steels and aluminium alloys. This type of hybrid joint should only be selected if there is an opportunity to fully cure the adhesive subsequently at high temperature. The joint can sustain much larger stresses before failure because of the homogeneous distribution of stresses in contrast to the use of spot welding on its own. There can also be improvements in total component stiffness as well as impact resistance if a tough adhesive is selected and used with the right components. Disadvantages are that the structure needs to be designed from the start with adhesive bonding in mind, which requires careful planning, good process control and specialised process equipment.

12.8.2 Folding/adhesive bonding

One of the oldest techniques of hybrid bonding in automotive engineering was the use of sealants or, if greater strength was needed, adhesives in folded components. These folds needed to be protected from exposure to moisture to prevent corrosion. Folded parts also do not always have the same mechanical properties in all directions. The use of structural adhesives both provides a seal and creates more uniform strength across the fold. In addition the NVH properties of the BIW are improved, due to the sealants taking up certain frequencies as well as providing overall tightening of the passenger compartment against (wind) noise. Through the use of adhesives, components can take higher loads and can absorb more energy in case of a crash. The stiffness of the total part is also increased, since less flexibility is present between the folded parts. For every folded part, the stress and strain in the folds have to be carefully taken into account, since with thinner materials, more buckling and rupture is expected. The benefit of folding with adhesive bonding is that the adhesive has a structural mechanical support from the folded joint until fully cured.

One of the potential disadvantages of strengthening folded joints with adhesives is that the adhesive, due to its lower viscosity, does not always have the full filling properties of a sealant initially, therefore potentially allowing moisture/water into the joint. If the joint is designed in the wrong way or the adhesive curing properties are not fully understood and controlled, curing might not be complete, resulting in lower bond strength. Washout of not fully cured adhesive could also occur in the coating process or pre-treatment of the BIW before coating. These problems can, however, be easily overcome by appropriate selection of adhesive material in conjunction with appropriate vehicle design to allow draining of the electro-coated paint as well as careful process control (e.g. to prevent moisture getting to the joint before curing is complete).

12.8.3 Adhesive bonding/(self-piercing) riveting

One of the more recent techniques used in aerospace engineering is hybrid bonding with adhesives and rivets. An adhesive that cures at room temperature (originally applied as a sealant) has been used alongside riveting. The curing of this adhesive at room temperature requires a relative short clamping time. The adhesive is used both as an aligning mechanism to speed up riveting as well as providing additional strength (see Fig. 12.2). The cured adhesive is then drilled through and the rivets installed. This method prevents chips

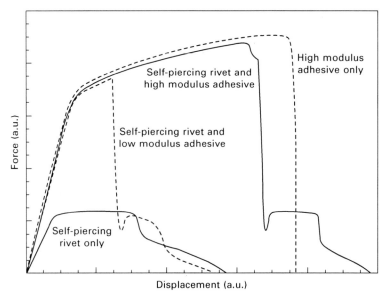

12.2 Schematic overview of the mechanical loading behaviour of hybrid self-piercing rivet and adhesive bonding combination as observed from various experiments.

from the drilling process ending up inside the structure. There is therefore no need to re-drill the holes, achieving a higher surface quality in each hole. This allows for a more precise alignment, reduces the waste caused by re-drilling holes that are outside the pre-set specifications and reduces costs by an increased speed of production. This technology is well suited to automotive engineering. The use of adhesives that cure at room temperature makes production easier, more flexible and saves costs. However, it should be note that the 'open' time (the time that the adhesive can still be applied after mixing of the components) plays a crucial role in the adhesive bonding process and needs to be carefully managed.

12.9   The effect of volume on joining technology

The cost of any joining technique, including the investment required, its reliability and flexibility, is a crucial consideration before its adaptation in the industry. Other considerations are the speed of the technique within the overall production line. The investment costs of the joining method are recovered once a certain predefined volume is met. As an example, a comparison of cost and speed for adhesive bonding with other techniques is shown in Table 12.2.

12.9.1 Low volume production

Spin-offs from other areas of transportation like the aerospace technology, developments in hand layup and fibre metal laminates (FMLs) have led to the introduction of new materials and techniques in the automotive sector. The integration of materials such as aluminium and glass or carbon fibre-reinforced polymer composites, as well as new joining techniques such as structural adhesive bonding in BIW construction, have proved to be cost-effective at low volumes. Fibre winding by robots to integrate all parts of a space frame may be a viable option in the premium sector, although the investment costs and R&D effort are considerable.

Table 12.2 Estimated costs and speed of spot welding, riveting, adhesive bonding and combinations thereof

Joining technique	Estimated cost (€/m)	Estimated joining speed (m/min)
Spot welding	1.5	0.8
Riveting	1.7	0.8
Adhesive bonding	0.4	8
Adhesive bonding and spot welding	1.4	0.7
Adhesive bonding and riveting	1.6	0.7

Values adapted from [2]

12.9.2 High volume production

Levels of automation of processes demand higher investment. Selection of the most appropriate joining technology will require consideration of many issues including material costs, production (facility) costs, speed of a production process, investment in production equipment and knowledge (people). The robustness of a production process and the equipment it needs is of great importance because unexpected breaks during production can run up production costs. The speed and productivity of a process can also be critical. As an example, in riveting, several production steps are necessary for the preparation of the needed rivet holes:

*   aligning the materials
*   placement of drilling masks
*   drilling itself
*   deburring and repositioning
*   riveting can take place.

The time and number of steps of varying complexity might call for a complete redesign of the original joint. In high-volume production, this requires very careful planning. Many alternatives might have to be compared before the final joint design is acceptable. In high volume production, failure rates of a particular joint type or design are a critical issue. There are many cases where cars have had to be recalled because of the possibility of failure of a certain material or joint. This makes it particularly important to evaluate joint designs and technologies before final selection and production.

12.10 Future trends

New and hybrid materials offer a good way of reducing the weight of a car. However, their advantages need to be weighed against the differing costs of raw materials. This needs to include the possibility of price rises if the demand for newer materials for automotive engineering grows, though it may be possible for larger manufacturers to buy materials in bulk at a reduced price. In the short term, new materials are more economically attractive at the premium price end of the market where higher costs can be more easily recouped. As an example, the benefits of high strength steels are starting to be accepted in the automotive industry, as always adopted first in higher priced, smaller series production cars.

With the pressing matter of materials scarcity, future trends may focus on more locally produced and recycled materials. Metallic materials have a great advantage as the technologies for recycling steels, aluminium and magnesium have matured and proved to be very efficient. A great deal of experience has been gained in using these recycled materials in new vehicles.

Addition of virgin material to recycled melts can actually improve material properties, whereas recycled polymeric materials without exception show considerable lower grade properties even when virgin material is added. Complex materials like carbon fibre-reinforced polymer composites are not easily recyclable.

Repair of automobiles is another issue which will dominate the automotive industry as materials get scarcer. If materials are scare, the cost of a new part or new car may far exceed the cost of repair, making repair a more attractive option than replacement. Plastics and FRP composite materials have one huge disadvantage here, since high quality (invisible) repairs are difficult, if not virtually impossible, in standard workshops and therefore full replacement may be required. Most of these composite constructions rely on the total lay-up (or winding) of the fibres and their interconnections. Repair will, therefore, always result in discontinuities, symmetry issues and, in general, a reduced performance. In the case of both thermoplastic and thermoset FRP composites, the aerospace industry has been developing designs and techniques to create lightweight but robust structures. A study into those methods could be well worth the effort in the long run for high-performance upper segment cars.

It is most likely, therefore, that joining techniques in the future will focus for the most part on metals, predominantly steels which, because of their low price and high performance, will retain a similar or even higher market share as light alloy prices rise. For low volume, premium price production, light metals offer advantages over carbon fibre in recyclability, repair and flexibility of production methods. Joining in the future will, therefore, focus on hybrid joining techniques for steel and other metallic materials. The relatively young technology of adhesive bonding will also increase given its range of benefits. As an example, using adhesive bonds based on cold curing adhesives as a basis for drilling and alignment for riveting would be useful for components where no heat input is allowed. The trend towards thinner but stronger materials continues. Joining methods will have to be continually adapted, with smaller HAZs and preferably no other heat-input than the high temperature baking of coatings.

12.11   References

[1] R. Overbey, Conference Board Session on Sustainability, Alcoa Primary Metals Development, June 2005.
[2] J. Vrenken, Oral presentation, Lijmen in de automotive industrie, Corus, 2009.
[3] P. Dhaeze, Integrated Composite Space Frame Design, Master Thesis, TU Delft, 2009.
[4] A. Fortunato, G. Cuccolini, A. Ascari, L. Orazi, G. Campana, G. Tani *Int. J. Mater. Form.*, Vol. 3 Suppl 1: 1131–1134, 2010.
[5] R. W. Messler, *Joining of Materials and Structures*, Elsevier, 2004.

Index

330

Lightning Source UK Ltd.
Milton Keynes UK
UKOW041248200712

196327UK00001B/101/P

9 781845 695613